Springer Series in Optical Sciences Volume 10

Edited by David L. MacAdam

Springer Series in Optical Sciences
Edited by David L. MacAdam

R. H. Kingston

Detection of Optical and Infrared Radiation

With 39 Figures

Springer-Verlag Berlin Heidelberg New York 1978

Dr. ROBERT H. KINGSTON
Massachusetts Institute of Technology, Lincoln Laboratory,
Lexington, MA 02173, USA

Dr. DAVID L. MACADAM
68 Hammond Street, Rochester, NY 14615, USA

ISBN 3-540-08617-X Springer-Verlag Berlin Heidelberg New York
ISBN 0-387-08617-X Springer-Verlag New York Heidelberg Berlin

Library of Congress Cataloging in Publication Data. Kingston, Robert Hildreth, 1928–. Detection of optical and infrared radiation. (Springer series in optical sciences; v. 10). Includes bibliographical references and index. 1. Optical detectors. 2. Infra-red detectors. I. Title. TA1522.K56 621.36 78-1486

Printed in Germany

Offset printing and bookbinding: Zechnersche Buchdruckerei, Speyer.
2153/3130–543210

Preface

This text treats the fundamentals of optical and infrared detection in terms of the behavior of the radiation field, the physical properties of the detector, and the statistical behavior of the detector output. Both incoherent and coherent detection are treated in a unified manner, after which selected applications are analyzed, following an analysis of atmospheric effects and signal statistics. The material was developed during a one-semester course at M.I.T. in 1975, revised and presented again in 1976 at Lincoln Laboratory, and rewritten for publication in 1977.

Chapter 1 reviews the derivation of Planck's thermal radiation law and also presents several fundamental concepts used throughout the text. These include the three thermal distribution laws (Boltzmann, Fermi-Dirac, Bose-Einstein), spontaneous and stimulated emission, and the definition and counting of electromagnetic modes of space. Chapter 2 defines and analyzes the perfect photon detector and calculates the ultimate sensitivity in the presence of thermal radiation. In Chapter 3, we turn from incoherent or power detection to coherent or heterodyne detection and use the concept of orthogonal spatial modes to explain the antenna theorem and the mixing theorem. Chapters 4 through 6 then present a detailed analysis of the sensitivity of vacuum and semiconductor detectors, including the effects of amplifier noise. Thermal detectors are then treated in Chapter 7, thermal-radiation-field fluctuations being derived using the mode concept and a semiclassical approach originated by HANBURY BROWN and TWISS (1957a). Chapter 8 again uses the spatial-mode concept, as well as the stimulated- and spontaneous-emission relationships to determine the advantages of laser preamplification prior to detection. Atmospheric limitations on detection efficiency are briefly reviewed in Chapter 9 with special emphasis on the effects of turbulence. Following a detailed discussion of the significance of signal-to-noise ratios in terms of detection probabilities in Chapter 10, radar, radiometry, and interferometry are used as a framework to demonstrate the application of the previous results.

I have not tried to be encyclopedic in the treatment, either in terms of complete references or a discussion of every type of detector. References are included for more detailed background or, in some cases, historical interest. The most common photon and thermal detectors are analyzed; more specialized devices may be easily understood by extension of the treatments in the text. As an example the charge-coupled photon detector (CCD) recently reviewed by MILTON (1977) is basically a semiconductor photodiode with integrated readout and amplification. A detailed breakdown of the whole family of detectors may be found in KRUSE (1977). Another valuable reference is the annotated collection of papers by HUDSON and HUDSON (1975).

As basic sources for the fundamentals of incoherent detection and noise, I found SMITH et al. (1968) and VAN DER ZIEL (1954) invaluable, the latter especially helpful for an understanding of detector-noise mechanisms. The final version of the work profited by suggestions and comments from my colleagues and students and I especially thank Robert J. Keyes, my co-worker for many years, for his help and advice on detector behavior. I am indebted to Marguerite Ampolo for typing the original lecture notes, to Debra Brown for preparation of the final manuscript, and to Robert Duggan for the illustrations.

Lexington, Massachusetts,
January 1978

R. H. KINGSTON

Contents

1. Thermal Radiation and Electromagnetic Modes

Before a discussion of the detection process, we first investigate the properties of the optical and infrared radiation in equilibrium with a cavity at temperature T. The results of this derivation are essential to an understanding of detection processes limited by thermal radiation from the vicinity of the detectors as well as from the background or other extraneous radiators near the desired signal source. The treatment also introduces the three forms of thermal statistics which will be used later in discussions of laser preamplification and thermal detectors. A up right final concept, introduced in this chapter and developed elsewhere in the text, is that of the allowed spatial modes of the electromagnetic field. The properties of these modes are key elements in the treatment of heterodyne detection and radiation field fluctuations.

1.1 The Nature of the Thermal Radiation Field

To treat the thermal radiation field we start with the applicable distribution statistics for "photons" and apply these statistics to the allowed modes of an electromagnetic field. First we consider the appropriate statistics, which are one of a class of three distinct types. These are Maxwell-Boltzmann, Fermi-Dirac, and Bose-Einstein. They are each the probability of occupancy of an allowed state and apply to different types of particles, as follows:

Maxwell-Boltzmann: Distinguishable particles, no exclusion principle. Limiting statistics for all particles for high energy or state occupancy much less than one,

$$f(E) = f_0 \mathrm{e}^{-E/kT}\,, \tag{1.1}$$

where E is the energy of the state and k = Boltzmann's constant = 1.38×10^{-23} J K^{-1}.

Fermi-Dirac: Indistinguishable particles obeying the exclusion principle, i.e., only one particle allowed per state (e.g., electrons),

$$f(E) = \frac{1}{\mathrm{e}^{(E-E_{\mathrm{F}})/kT} + 1}\,. \tag{1.2}$$

Bose-Einstein statistics: Indistinguishable particles and unlimited state occupancy,

$$f(E) = \frac{1}{e^{(E-E_B)/kT} - 1} \,. \tag{1.3}$$

For the case of the modes of an electromagnetic field or "photons", the number of "particles" is unlimited and the distribution becomes

$$f(E) = \frac{1}{e^{E/kT} - 1} \tag{1.4}$$

and is usually referred to as "photon statistics". In both the Fermi-Dirac and Bose-Einstein statistics, the fixed energy parameter E_F or E_B is adjusted so that the total distribution in the system adds up to the total number of available particles. In the case of photons, the number is not limited and E_B becomes zero. The derivation of these statistics may be found in standard texts (see, for example, *Reif,* 1965) and is a subject in itself. We therefore shall not justify them at this time but later shall derive some relationships based on radiation theory which make them self-consistent and at least plausible.

1.2 Derivation of Planck's Radiation Law

We start with the following theorem:
In a large cavity (dimensions large compared to a wavelength), each allowed electromagnetic mode of frequency ν has energy $E = h\nu$, and the number of the modes excited is determined by the Bose-Einstein statistics applicable for photons. The temperature is determined by the temperature of the wall or of any absorbing particle in the cavity, under equilibrium conditions.

Consider a large cavity with slightly lossy walls, which for convenience in counting modes will be a parallelopiped of dimensions L_x, L_y, and L_z. To count the modes as a function of frequency, we write the standing-wave solution to Maxwell's equation as

$$E = E_0 \sin k_x x \sin k_y y \sin k_z z \sin 2\pi\nu t$$

subject to the boundary conditions that

$$k_x L_x = n\pi, \text{ etc} \,.$$

We also note that Maxwell's equation,

$$\nabla^2 E = (1/c^2)\partial^2 E/\partial t^2$$

requires that

$$k_x^2 + k_y^2 + k_z^2 = 4\pi\nu^2/c^2 \,.$$

Using these relationships, we can construct the distribution of allowed modes in k space as an array of points occurring at $k_x = n\pi/L_x$, $k_y = n\pi L_y$, etc., where n is an integer 1 or greater. The density of points in k space, as seen from Fig. 1.1, becomes

$$\rho_k = L_x L_y L_z/\pi^3 = V/\pi^3 ,$$

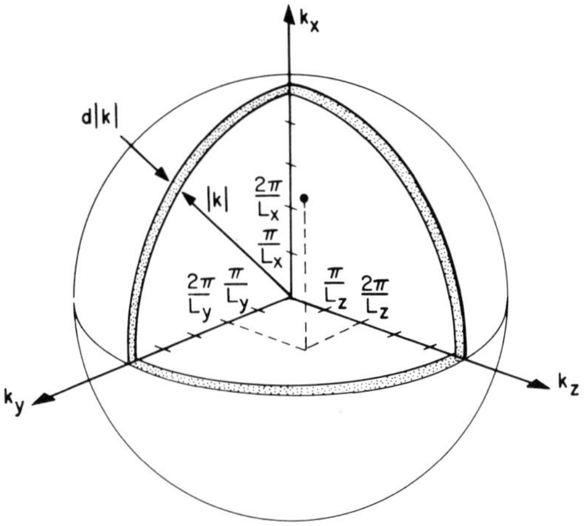

Fig. 1.1. Mode distribution in k space

where V is the volume of the cavity. To determine the density of modes versus frequency, we note that

$$|k| = \frac{2\pi\nu}{c} ,$$

as shown in Fig. 1.1. Therefore, a spherical shell in the first octant of k space contains points that represent all modes at frequency $\nu = |k|c/2\pi$. We then may solve directly for the number of modes dN in a frequency range $d\nu$

$$\begin{aligned} dN &= \frac{\pi}{2} \cdot \rho_k \, k^2 dk \\ &= \left(\frac{\pi}{2}\right)\left(\frac{V}{\pi^3}\right)\frac{4\pi^2\nu^2}{c^2} \cdot \frac{2\pi d\nu}{c} \qquad (1.5) \\ &= \frac{4\pi V}{c^3} \cdot \nu^2 d\nu . \end{aligned}$$

To calculate the energy per unit frequency interval, we must take into account the allowed occupancy of the modes. This is given by (1.4) and also must include the fact that two separate orthogonal polarizations are allowed for each mode. Thus

$$du = dE/V = 2f\!\!\!/\, h\nu dN/V \tag{1.6}$$

and the final result is the energy per unit volume per unit frequency range,

$$du_\nu = \frac{8\pi h}{c^3} \cdot \frac{\nu^3 d\nu}{(e^{h\nu/kT} - 1)}, \tag{1.7}$$

where h = Planck's constant = 6.6×10^{-34} Js.

We now ask what is the power density that strikes the cavity wall or crosses any small surface area within the cavity. We know that the energy density arises from forward and backward travelling waves that produce standing wave modes, and that the waves and the energy are moving at the velocity of light c. If the surface has area A, the power striking it at an angle θ from the normal is given by

$$dP_\nu = \frac{c}{2}\, du_\nu\, \frac{A \cos\theta\, d\Omega}{2\pi}, \tag{1.8}$$

where, as shown in Fig. 1.2, the power density that flows in a small solid angle $d\Omega$ is one-half the energy density multiplied by the velocity of light, reduced by the ratio of the solid angle to a full hemisphere. The power collected by the surface is proportional to $A \cos\theta$ because of the angle of incidence of the flux.

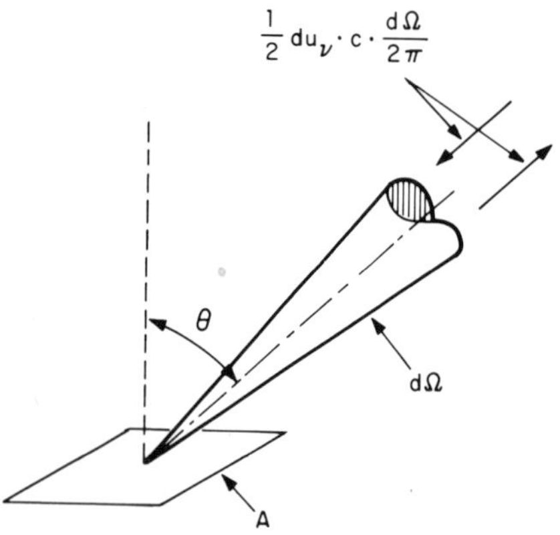

Fig. 1.2. Power flow in thermal-radiation field

This argument assumes that the radiation flow is isotropic, which could be proved by use of the cavity treatment but which we shall show later in a more general proof. Integrating over the hemisphere, with

$$d\Omega = 2\pi \sin\theta\, d\theta$$

yields

$$P_\nu = \frac{c}{2} du_\nu \int_0^\pi \frac{A \cos\theta \, 2\pi \sin\theta \, d\theta}{2\pi} = \frac{c}{4} du_\nu A \tag{1.9}$$

and

$$dI_\nu = \frac{2\pi h}{c^2} \frac{\nu^3 d\nu}{(e^{h\nu/kT} - 1)}, \tag{1.10}$$

where I_ν is defined as the irradiance or the total power per unit surface area. This last expression is the commonly used form of Planck's radiation law. The distribution for $T = 300$ K is plotted in Fig. 1.3 with ν^*, the reciprocal wavelength (wave numbers) as the variable.

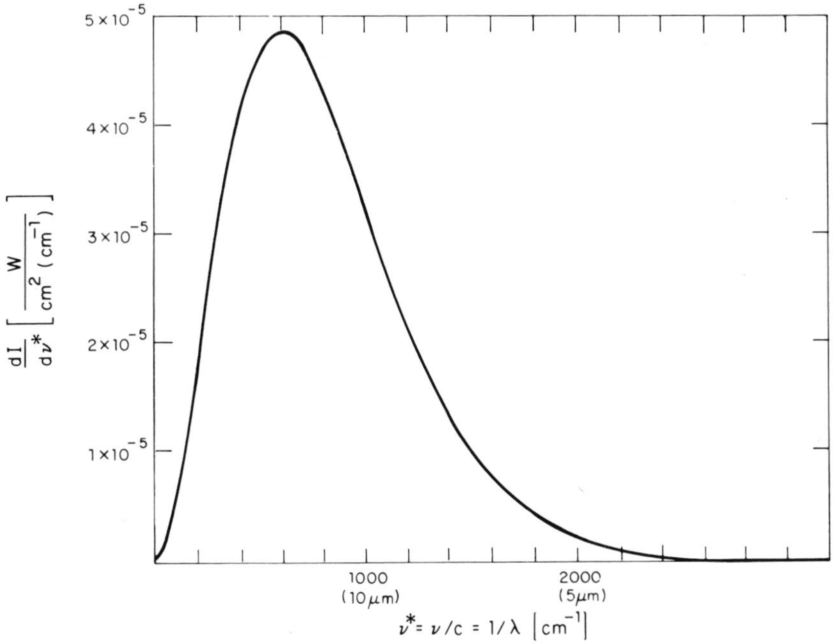

Fig. 1.3. Planck distribution for $T = 300$ K. For any temperature T, multiply the abscissa, ν^*, by $T/300$, and the ordinate $dI/d\nu^*$ by $(T/300)^3$

1.3 General Properties of Blackbody Radiation

In our derivation, we assumed an ideal cavity with slightly lossy walls in order to derive the radiation law. We now generalize and introduce the concepts of emissivity ε and absorptivity α. By use of a general principle known as detailed balance we may show that

1) The emissivity of a body is equal to its absorptivity at all frequencies and at all incidence angles.
2) The radiation is isotropic and independent of position in the cavity.
3) The radiation law is true for *any* enclosed region of arbitrary shape, provided that the temperature of all walls and enclosed objects is the same.

We first insert in the cavity a small particle that absorbs completely at all frequencies. If the particle is at the same temperature as the walls, then the energy it absorbs should equal the energy it radiates, otherwise its temperature would change. We imagine the particle to move throughout the cavity. Because the energy it radiates is a function of temperature alone, the energy incident on it must be independent of position, otherwise the temperature of the particle would change, in violation of the equilibrium requirement. Therefore, the radiation field is independent of position. We next assume that the particle has emissivity less than unity. In this case, the particle radiates less energy than the energy impinges on it. To maintain constant temperature, its reflectivity must balance its decreased emission. We therefore require that the emissivity be equal to (1 − reflectivity), which is equal to the absorptivity. In a similar manner, using arbitrary shields or frequency filters about the particle, we can show that the radiation in the cavity is isotropic and that the equality of emissivity to absorptivity holds at all frequencies and angles of incidence upon a surface. Finally, we imagine our sample cavity to be coupled through a small hole to a second cavity of arbitrary shape and with walls of any emissivity, all maintained at the equilibrium temperature T. By inserting frequency, angular, or polarization filters between the two cavities, we may show that the radiation in both cavities must have the same frequency, angular, and polarization behavior; otherwise the second cavity would absorb or lose energy, again violating the equal-temperature requirement. For further details the reader is referred to *Reif* (1965).

We now deduce two important consequences of the foregoing arguments. These are, first, that a surface with emissivity ε and temperature T emits energy, *even in the absence* of incident radiation, at a rate that is equal to the rate at which it absorbs energy from the radiation that is incident on it in a cavity that is at the same temperature. If we define the radiance in $\mathrm{Wm^{-2}}\cdot$steradian, $H_{\nu,\Omega}$, then the radiance from a surface of area A is

$$dH_{\nu,\Omega} = \varepsilon(\nu,\theta) \cdot \frac{2h}{c^2} \frac{\nu^3 d\nu \cos\theta}{(e^{h\nu/kT} - 1)} \tag{1.11}$$

and the total emitted power for isotropic ε is

$$dP_\nu = \varepsilon \cdot \frac{2\pi h}{c^2} \frac{A\nu^3 d\nu}{(e^{h\nu/kT} - 1)}, \tag{1.12}$$

which, when integrated over frequency, yields the Stefan-Boltzmann equation.

An extension of this rule may be applied to a partially absorbing medium distributed in a small volume of the cavity. In this case, we surround the absorbing or partially transparent medium by an imaginary boundary and, considering one direction of energy flow through the bounded region, require that the net flow into one side equals the net flow out. If the medium is to maintain constant temperature, then the radiation absorbed along the path must equal the radiation emitted. It then follows that the medium emits in the forward *and* backward directions with an effective emissivity equal to the absorptivity or fractional power loss along the path. If we remove this absorbing medium from its cavity but maintain its temperature, it will appear against a radiationless background as an emitter that radiates according to the same laws as an emissive surface at the same temperature.

1.4 A Plausibility Test of the Planck Distribution

Although we presented the three statistical distributions without proof, it is instructive to show the interaction between two of them to indicate their physical plausibility. We do this by calculating the interaction between a simple two-level Maxwell-Boltzmann system and the radiation field inside a blackbody cavity. This treatment was used in a reverse manner by Einstein to derive the Bose-Einstein statistics. We start by defining the Einstein A and B coefficients for atomic transitions, as follows. In a two-level system with low occupancy, the probability of transition from a higher energy state, level 2, is given by

$$P(\text{emission}) = (A + Bu)\,f_2\,, \tag{1.13}$$

where A represents the spontaneous-emission term and Bu is the induced emission, which is proportional to the energy density in the radiation field. f_2 is the Maxwell-Boltzmann factor for the upper state, $\exp(-E_2/kT)$. The absorption rate is

$$P(\text{absorption}) = Bu\,f_1\,; \tag{1.14}$$

if the atomic system is in equilibrium with the radiation field, the two rates should be equal. Thus we have

$$A/B = u(f_1 - f_2)/(f_2) = u(e^{-(E_2-E_1)/kT} - 1)\,.$$

Substituting the Planck radiation density for u yields

$$A/B = 8\pi h\nu^3/c^3\,,$$

which we note is completely independent of temperature. This is satisfying physically, because it indicates that the ratio of spontaneous emission *power* to induced power is proportional to ν^4, as would be expected from classical theory.

1.5 Numerical Constants and Typical Values

The physical constants used in typical calculations of the radiation field are

$$h = 6.6 \times 10^{-34}\ \mathrm{Js}$$
$$k = 1.38 \times 10^{-23}\ \mathrm{JK}^{-1}$$
$$c = 3 \times 10^{8}\ \mathrm{ms}^{-1}$$
$$e = 1.6 \times 10^{-19}\ \mathrm{C}\,.$$

During most calculations, it is more convenient to express photon energies and thermal energy in eV, which are defined as the voltage difference through which an electron would have to fall to gain the same energy. Thus, if the energy is 1 eV, the energy in joules is 1 V times the electron charge, or 1.6×10^{-19} J. The photon energy is

$$E = h\nu = hc/\lambda$$

which, expressed in eV is

$$(h\nu)_{\mathrm{eV}} = hc/e\lambda = 1.24/\lambda\,,$$

where the wavelength λ is expressed in μm. Similarly, the thermal energy in eV is

$$(kT)_{\mathrm{eV}} = \frac{kT}{e} = 0.026\,(T/300)\,,$$

where T is expressed in K.

On the basis of these relationships, it is instructive to calculate the value of the Bose-Einstein occupancy factor for typical temperatures and wavelengths. The result for $T = 300$ K, and for a wavelength of 1 μm (just into the infrared from the visible), is

$$f = \frac{1}{\mathrm{e}^{h\nu/kT} - 1} = \frac{1}{\mathrm{e}^{48} - 1} \simeq 10^{-21}\,.$$

For 10 μm, about the longest infrared wavelength propagated through the atmosphere, the factor is

$$\mathscr{f} = \frac{1}{e^{4.8} - 1} = \frac{1}{120}\,.$$

Thus, in either of these cases, the state occupancy is much less than unity; we shall use the following approximation in many of the treatments to follow:

$$\mathscr{f} = \frac{1}{e^{h\nu/kT} - 1} = e^{-h\nu/kT} \qquad h\nu \gg kT\,.$$

Problems

1.1 Derive the Stefan-Boltzmann law by integrating over the Planck distribution. $I = \varepsilon\sigma T^4$. Show that $\sigma = (2\pi^5 k^4)/(15c^2h^3) = 5.67 \times 10^{-8}$ $\mathrm{Wm^{-2}\ deg^4}$. [The integral $\int_0^\infty (x^3 dx)/(e^x - 1) = \pi^4/15$].

1.2 A surface is in equilibrium with a blackbody field of temperature 300K.

a) What is the total irradiance on the surface?

b) At what wavelength λ is the irradiance per unit frequency a maximum? Check your answer with Fig. 1.3.

1.3 The sun's diameter subtends an angle of 9.3 milliradians as seen from the earth. If the sun is a perfect blackbody at temperature $T = 5800$ K, find the solar constant, the irradiance at the top of the earth's atmosphere.

1.4 An optical receiving element of area A has diffraction-limited beamwidth given by $\Omega = \lambda^2/A$, where Ω is the solid angle of the beam. Find the power per unit frequency interval incident on the receiver within this solid angle Ω. Find the value for the limiting case where $h\nu \ll kT$.

1.5 In Section 1.3, we derived the ratio of the Einstein A and B coefficients by placing a Maxwell-Boltzmann distribution in equilibrium with the radiation field. Repeat the calculation using a Fermi-Dirac system, again in equilibrium with the radiation. Note that the transition probability is now proportional to the occupancy of the initial state times the "emptiness" of the final state. You will thus have terms in $\mathscr{f}_2(1 - \mathscr{f}_1)$, etc.

2. The Ideal Photon Detector

We now treat the optical and infrared detection process by postulating the ideal photon detector. This is a device that samples the incident radiation and produces a current proportional to the total power incident upon the the detector surface. We first consider the fundamental principles of detection, then the noise associated with the detection process, and, finally, analyze two limiting forms of detection, signal-noise limited and background-noise limited. In the latter case, we shall use the results of Chapter 1 to find the limiting sensitivity in the presence of thermal radiation. The treatment in this section ignores noise from the following amplifier stages as well as noise sources peculiar to real devices; however, the results set the absolute limits of sensitivity for a device that exhibits ideal behavior.

2.1 Event Probability and the Poisson Distribution

Here we start with a fundamental theorem of photodetection that is the basis of the whole treatment of detection theory. The theorem is: If radiation of constant power P is incident upon an ideal photon detector, then electrons will be produced at an average rate given by

$$\bar{r} = \eta P/h\nu$$

where η is defined as the quantum efficiency, that is, the fraction of the incident power effective in producing the emitted electrons. A second and most important part of the theorem is that the electron-emission events are randomly distributed in time. Because of the quantum nature of radiation, each photoevent or electron results from the extraction or loss of one "photon" or $h\nu$ of energy from the incident field. If the incident radiation is time varying, the *average* rate will vary with time in the same manner. For constant $\bar{r}$, such a random process obeys Poisson statistics, which state that the probability of the emission of k electrons in a measurement interval τ is

$$p(k, \tau) = \frac{(\bar{r}\tau)^k e^{-\bar{r}\tau}}{k!} . \tag{2.1}$$

A derivation of this distribution may be found in *Davenport* and *Root* (1958). In the case of time-varying power and thus $\bar{r}$, the Poisson distribution is still valid, provided that the sampling interval τ is short compared to any characteristic period of the power variation. At this stage we shall not use the distribution function explicitly. That subject is to be considered in Chapter 10 on detection and false-alarm probabilities. We shall use a particular property of the distribution, however, which is the mean-square fluctuation of the number of events in a fixed time period when averaged over many observations. This expression, familiar from simple statistics, is

$$\overline{(n-\bar{n})^2} = \overline{n^2} - \overline{2n\bar{n}} + \bar{n}^2 = \overline{n^2} - \bar{n}^2 = \bar{n} \,. \tag{2.2}$$

In the special case of thermal radiation with $h\nu$ comparable to or less than kT, even "constant" thermal power will not produce the above distribution. Most texts (e.g., *Ross*, 1966) treat the resultant output current fluctuation in terms of the fluctuation in the Bose-Einstein occupancy factor. We shall show in Chapter 8 that the same result may be obtained by treating the thermal radiation power as a fluctuating quantity in a classical manner; the fluctuation occurrs at frequencies from zero to the full spectral bandwidth.

2.2 Noise in the Detection Process

Because the ideal photon detector produces a sequence of narrow equal-amplitude pulses, the optimum way to process the information is by counting individual pulses. In most optical and infrared detection systems, however, either the pulse rate is so high that the pulses overlap and become indistinguishable or competing noise in the detector or amplifier circuits makes an individual pulse nondistinguishable from the noise fluctuations. As shown in Fig. 2.1, current pulses from a photodetector will always have finite pulsewidth τ_p because of the limited response time of the detector. Alternatively, the post-detection filtering used to match the detector system to the signal bandwidth sets a minimum pulsewidth limit on the system output. For cases in which the pulsewidth is short compared to the average pulse-repetition period, and the pulse amplitude is well above circuit noise, pulse-counting techniques (Chap. 10) are most efficient. In the other extreme, when there are many pulses per sampling interval or filter response time, the noise associated with the random pulse distribution becomes a small fluctuation about the dc mean. By the central-limit theorem (see *Davenport* and *Root,* 1958; *Bracewell*, 1965) the fluctuation becomes Gaussian, as shown in Fig. 2.1, where we have plotted the probability distribution function for the output current. Later, we shall discuss the Gaussian distribution in much more detail, but for the present we are concerned with the mean-square fluctuation or "noise" compared with the power produced by the incident signal. The last graph in Fig. 2.1 depicts the second essential descriptor for the noise, its power spectral density or, more simply, the mean-square fluctuation per unit frequency range as a function of frequency. This quantity,

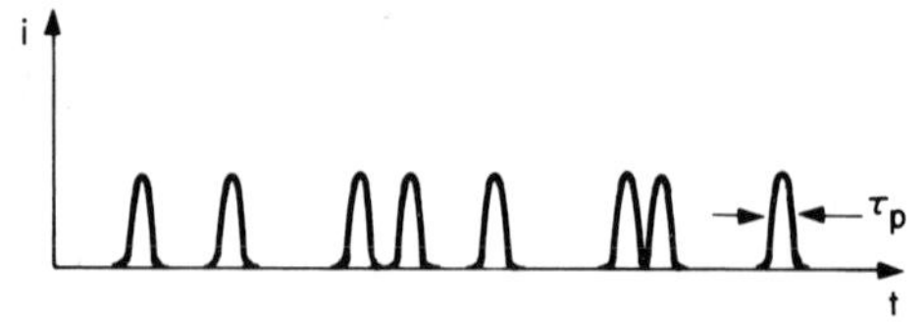

(a) Waveform for low count rate. τ_p determined by detector or filter response

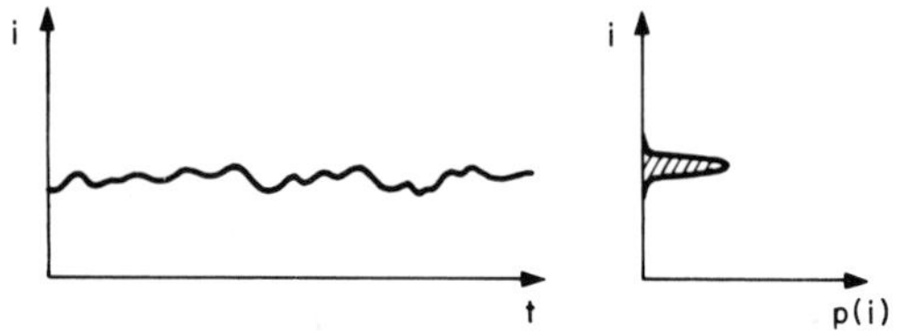

(b) Waveform for large number of pulses per unit response time and resultant probability density function

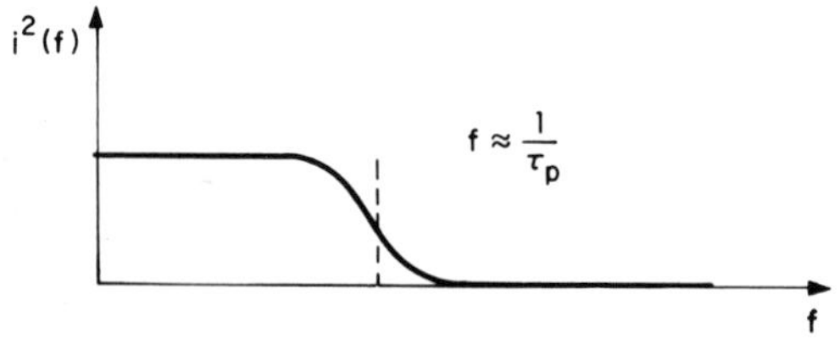

(c) The power spectral density function for the noise current

Fig. 2.1a-c. Photodetector-current characteristics

combined with the detection-system frequency response, determines the ultimate noise output from the system. Taking the ratio of the output signal power to the noise power yields the power signal-to-noise ratio, $(S/N)_P$, which we shall use as a uniform figure of merit for detectors throughout the text. The required $(S/N)_P$ for any particular application is dependent upon the system and the desired performance. These matters are discussed in detail in Chapters 10 and 11.

We now proceed to two separate derivations of the expected current fluctuation in the output of the detector. The first derivation is somewhat qualitative but presents the physical meaning of the noise current, whereas the second treatment is rigorous and is based upon fundamental noise theory. Consider a filter coupled to the output of the detector that is excited by the narrow current pulses associated with each electron. We shall choose, for simplicity, a low-pass filter that has an integration time of τ. What we wish to determine is the mean-square current fluctuation at the output of the filter; using the effective filter bandwidth, we then can determine the spectral density of the mean-square noise current. We proceed by noting that the filter has a memory τ such that the net charge in the filter at any time is equal to the number of electrons that enter it during any time interval τ. The current measured during the time interval τ is

$$i = ne/\tau\,,$$

whereas the average current is

$$I = \bar{i} = \bar{n}e/\tau\,.$$

The mean-square current fluctuation, averaged over many independent measurement times, all of length τ is, from (2.2)

$$\overline{i_N^2} = \overline{(i - \bar{i})^2} = \frac{e^2}{\tau^2}\,\overline{(n - \bar{n})^2} = \frac{e^2}{\tau^2}\,\bar{n}$$

and the resultant mean-square noise current is

$$\overline{i_N^2} = \frac{e^2}{\tau^2}\,\bar{n} = \frac{e^2}{\tau^2}\cdot\frac{I\tau}{e} = \frac{eI}{\tau}\,. \tag{2.3}$$

It should be noted that the width of each pulse, τ_p, is much less than the sampling time τ. We must now ask what the effective bandwidth of the filter is when averaged with respect to power, because the mean-square current is a measurement of the power delivered from the filter. The bandwidth for an integrating filter with sampling time τ is given by (*Bracewell*, 1965)

$$B_{\mathrm{eff}} = \int_0^\infty \left(\frac{\sin\pi f\tau}{\pi f\tau}\right)^2 df = \frac{1}{2\tau}\,, \tag{2.4}$$

where we have integrated over the power response of the filter. (We shall use f for frequency of the detected current and ν for optical or infrared frequency). The resultant mean-square noise current becomes

$$i_N^2 = \frac{eI}{\tau} = 2eIB\,, \tag{2.5}$$

which is the well-known "shot-noise" expression used in vacuum-diode theory. A more rigorous treatment, based on the autocorrelation function (*Davenport* and *Root*, 1958, p. 122) yields

$$\overline{i_N^2}(f) = 2\bar{r}\,\,|i(\omega)|^2\,, \tag{2.6}$$

where $\overline{i_N^2}(f)$ is the spectral density of the mean-square noise current, and the term within the absolute magnitude signs is the Fourier transform of the current pulse, given by

$$i(\omega) = \int_{-\infty}^{+\infty} i(t) \mathrm{e}^{-\mathrm{i}\omega t}\, dt\,. \tag{2.7}$$

This derivation allows us to include the shape of the current pulse as determined by the detector response. If we consider a square pulse of total charge e and pulsewidth τ_p we obtain

$$i(\omega) = \int_0^{\tau_\mathrm{p}} \frac{e}{\tau_\mathrm{p}} \mathrm{e}^{-\mathrm{i}\omega t}\, dt = \frac{e}{\tau_\mathrm{p}} \left(\frac{1 - \mathrm{e}^{-\mathrm{i}\omega\tau_\mathrm{p}}}{\mathrm{i}\omega} \right), \tag{2.8}$$

which, for τ_p approaching zero, becomes

$$i(\omega) = e\,.$$

The final expression for the spectral density is then

$$\overline{i_\mathrm{N}^2}(f) = 2\bar{r}e^2 = 2\left(\frac{I}{e}\right) e^2 = 2eI\,, \tag{2.9}$$

as previously derived. We should note, incidentally, that the frequency response for the signal power is the same as the noise spectrum, so that the ratio of signal-to-noise power is a constant, if no additional noise is added in successive stages. We shall later use this more rigorous treatment in calculating both the spectral noise density and frequency response of real detectors.

2.3 Signal-Noise-Limited Detection

On the basis of our description of the noise mechanism in an ideal detector, we now consider the output signal-to-noise ratio for the case of signal radiation alone, without any extraneous radiation present. It is applicable when the detector is well filtered and is looking at a narrow-spectral source or when competing background radiation or scatter is negligible. We now assume that the detector is followed by a filter of bandwidth B, where we have chosen the bandwidth to match the waveform or frequency spread of the incoming signal-power envelope. In this case, we may write the signal current from the detector as

$$i_\mathrm{S} = \frac{e\eta P_\mathrm{S}}{h\nu}, \tag{2.10}$$

and the mean-square noise current becomes

$$\overline{i_\mathrm{N}^2} = 2\, e i_\mathrm{S} B = \frac{2e^2\eta P_\mathrm{S} B}{h\nu}\,. \tag{2.11}$$

The ratio of the signal power to the noise power at the output of the filter is then just the ratio of the squared currents, namely

$$\left(\frac{S}{N}\right)_{\mathrm{P}} = \frac{i_{\mathrm{S}}^2}{i_{\mathrm{N}}^2} = \frac{e^2\eta^2 P_{\mathrm{S}}^2}{(h\nu)^2} \cdot \frac{h\nu}{2e^2\eta P_{\mathrm{S}} B} = \frac{\eta P_{\mathrm{S}}}{2h\nu B}. \tag{2.12}$$

We may now introduce a common descriptor in optical detectors, which is the noise equivalent power (*NEP*). For the signal-noise-limited case, this is

$$(NEP)_{\mathrm{SL}} = \frac{2h\nu B}{\eta}, \tag{2.13}$$

which is the signal power required to yield $(S/N)_{\mathrm{P}} = 1$. The physical meaning of this expression may be seen more clearly if we use the relation between the filter bandwidth B and the effective time constant of the filter. Because $B = 1/2\tau$ from (2.4), the *NEP* for unit quantum efficiency is

$$(NEP)_{\mathrm{SL}} = \frac{h\nu}{\tau}, \tag{2.14}$$

which states that, *on the average*, one signal photon may be detected per unit measurement time. We emphasize that this is an average detection capability, because the Poisson statistics of the events must be examined for the actual *probability* of detection during a finite measurement interval. These matters will be considered in Chapter 10.

2.4 Background-Noise-Limited Detection

We now consider the detector in the presence of signal as well as background radiation. This latter may be due to solar scatter from clouds in the visible or thermal radiation from the surroundings in the case of an infrared detector. If the background is assumed to be constant, or more specifically, if the fluctuation of the background intensity is not at frequencies that fall within the bandwidth of the post-detection filter, the mean-square fluctuation of the output current from the filter is

$$\overline{i_{\mathrm{N}}^2} = 2e\bar{i}B = \frac{2\eta e^2(P_{\mathrm{S}} + P_{\mathrm{B}})\,B}{h\nu}, \tag{2.15}$$

whereas the current associated with the desired signal is, as previously,

$$i_{\mathrm{S}} = \frac{\eta e P_{\mathrm{S}}}{h\nu}.$$

The power signal-to-noise ratio is then

$$\left(\frac{S}{N}\right)_P = \frac{\eta^2 e^2 P_S^2}{(h\nu)^2} \times \frac{h\nu}{2\eta e^2(P_S + P_B)} = \frac{\eta P_S^2}{2h\nu B(P_S + P_B)}. \tag{2.16}$$

From this expression, we may write the noise-equivalent power for background-limited detection as

$$(NEP)_{BL} = \sqrt{\frac{2h\nu B(P_S + P_B)}{\eta}} \underset{P_S \ll P_B}{=} \sqrt{\frac{2h\nu BP_B}{\eta}}. \tag{2.17}$$

Here we should note a peculiar distinction of background-limited incoherent optical and infrared detectors, namely that the minimum discernible power is proportional to the *square root* of the bandwidth rather than to the bandwidth itself, as is the case in rf and microwave systems. This behavior results from the simple fact that the *output* power from a photon detector, or from any incoherent radiation detector, is proportional to the *square* of the input power. Thus for example, if the bandwidth is doubled, the output noise power is doubled but only a square-root-of-two increase of signal is required to obtain the same output signal-to-noise ratio. Although we shall not consider additional noise sources in the detector or in the amplifier at this time, we point out that the general behaviors of the $(S/N)_P$ and the *NEP* are the same in most real cases, because these excess noises generally result in an output noise power proportional to bandwidth (Johnson noise in the load resistor, for example), and behave in the same manner as the background-radiation-induced noise.

2.5 *NEP* and *D** in the Presence of Thermal Background

We now define two terms often used to describe optical and infrared detectors, namely D and D^*. D, the detectivity, is simply the inverse of *NEP* and is usually specified for a 1 Hz bandwidth, if not otherwise indicated. It is also often specified at a given wavelength, "chopping" or modulation frequency, and source radiation temperature. We shall not pursue the definition in detail other than to point out the obvious: detectivity is a figure of merit that increases as sensitivity improves. The second term, D^*, is called the *specific* detectivity (*Jones*, 1959) and is the value of D normalized to a detector area of 1 cm^2 when B is specifically designated to be 1 Hz. The logic of this measure may be understood as follows. Consider the *NEP* for a background-limited detector in the presence of background irradiance I. Then the quantity D may be written

$$D = \frac{1}{(NEP)_{BL}} = \sqrt{\frac{\eta}{2h\nu BP_B}} = \sqrt{\frac{\eta}{2h\nu I}} \cdot \frac{1}{\sqrt{AB}}, \tag{2.18}$$

and D^* is defined as

$$D^* = \frac{\sqrt{AB}}{(NEP)_{BL}} = \sqrt{\frac{\eta}{2h\nu I}}, \tag{2.19}$$

where A is in cm^2 and B is in Hz. Thus $D = D^*$ for a 1 cm^2 detector and 1 Hz bandwidth. The significant point is that the value of D^* for a truly background-limited detector is independent of the size of the detector and the $(NEP)_{BL}$ for any detector can be derived if the background irradiance and detector size are specified. A further advantage of the D^* figure of merit is that it applies to any detector for which the mean-square noise current is directly proportional to the *area* of the detector. This is universally true for the background-limited case and is sometimes, but not always, the case for detectors in general. An example for which this relation holds is the dark or thermionic current in a vacuum photodiode, which is proportional to detector area and produces shot noise in the same manner as the background-induced current. In any event, for an application in which the background radiation is the limiting noise source, the value of D^* under the specified conditions sets the absolute upper limit on the sensitivity of the system.

2.6 An Illustrative Example of Background-Limited Detection

We now use the foregoing results to calculate the sensitivity of a typical background-limited detection system, specifically a scanning thermal sensor that operates in the 10 μm wavelength region. This instrument might be used for infrared ground surveillance or for thermography in medical applications. We shall assume an angular resolution of 1 milliradian or a field-of-view matched to the detector size of 10^{-6} steradians. Because the peak of 300 K blackbody radiation is near 10 μm, we shall assume use of a narrowband infrared filter at this wavelength with 10% bandwidth. It is essential that this filter be cooled well below 300K, so that the detector sees only radiation from the entrance aperture and no radiation from the filter at wavelengths outside the filter bandwidth. A schematic of the system is shown in Fig. 2.2; the object field is a surface that has small temperature variations about an average value, $T = 300$ K. The other significant parameters are the emissivity of the object, which we assume to be unity, and the aperture area, which we take to be 100 cm^2, or an aperture diameter of about 4 in. We now ask, what is the limiting temperature variation that may be measured with the system. In particular, we wish to find the quantity $NE\Delta T$, the noise-equivalent temperature change, that is, the temperature change across the target that would produce a signal-to-noise ratio of unity in the detector output. Because we wish to scan the whole system rapidly to obtain an image, we shall choose a bandwidth B of 1 MHz, allowing frame times and

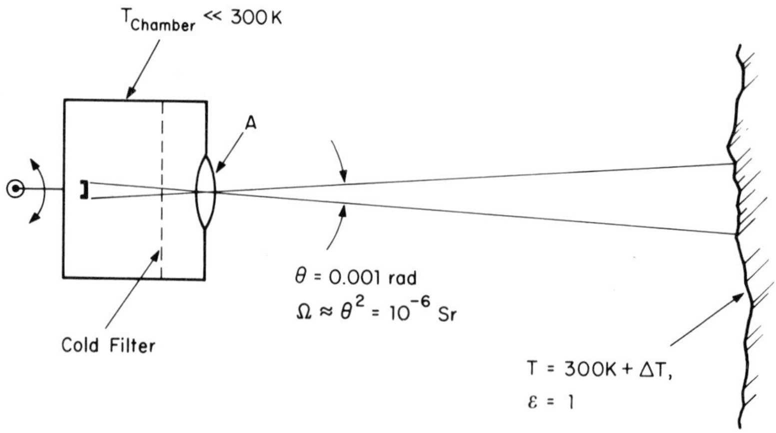

Fig. 2.2. Fast-scanning thermal imaging system

picture elements comparable with those of a TV system. We first calculate the background power that strikes the detector, by determining the amount of thermal radiation within the field of view Ω that strikes the aperture of area A at normal incidence. We note that, if the detector area exactly matches the field of view, then all power within this acceptance angle falls on the detector and the result is independent of the focal length; the latter determines only the appropriate size for the detector. Now from (1.11), we may write the irradiance per unit solid angle normal to the aperture as

$$dI_{\nu,\Omega} = dH_{\nu,\Omega} = \frac{2h\nu^3 d\nu}{c^2(e^{h\nu/kT} - 1)} \quad for\ \varepsilon = 1\ .$$

Taking into account the small angle Ω and approximating the integral over ν by the product of the center frequency irradiance times $\Delta\nu$, we obtain

$$P_B = A\Omega\,\frac{2h\nu^3\Delta\nu}{c^2(e^{h\nu/kT} - 1)} \simeq A\Omega\,\frac{2h\nu^3\Delta\nu}{c^2}\,e^{-h\nu/kT}\,,$$

where we have also used the high-energy approximation for the Bose-Einstein factor for $h\nu \gg kT$. For a change of temperature ΔT, the resultant signal power is

$$P_S = \Delta P_B = (dP_B/dT)\,\Delta T$$

and differentiation of the P_B equation with respect to temperature yields

$$dP_B/dT = \frac{P_B}{T}\left(\frac{h\nu}{KT}\right).$$

Thus the value of $NE\Delta T$ is

$$NE\Delta T = \frac{(NEP)_{BL}}{dP_B/dT} = \frac{T}{P_B}\left(\frac{kT}{h\nu}\right)\sqrt{\frac{2h\nu B \cdot P_B}{\eta}}\,.$$

We first calculate the background power P_B,

$$P_B = A\Omega \frac{2h\nu^3 \Delta\nu}{c^2}\, e^{-h\nu/kT} = A\Omega \frac{2(h\nu)\nu^3}{c^2} \cdot \frac{\Delta\nu}{\nu}\, e^{-h\nu/kT}$$

$$= (10^{-2})(10^{-6})\, \frac{2[(0.124)\,(1.6 \times 10^{-19})]\,(3 \times 10^{13})^3\,(0.1)}{(3 \times 10^8)^2} \cdot \frac{1}{120}$$

$$= 10^{-7}\ \mathrm{W}\,,$$

where we have used the constants of Section 1.5. Then, solving for $NE\Delta T$, with $\eta = 1$,

$$NE\Delta T = \frac{T}{P_B}\left(\frac{kT}{h\nu}\right)\sqrt{\frac{2h\nu B \cdot P_B}{h\nu}} = T\left(\frac{kT}{h\nu}\right)\sqrt{\frac{2h\nu B}{\eta P_B}}$$

$$= 300\left[\frac{1}{4.8}\right]\sqrt{\frac{2(0.124)\,(1.6 \times 10^{-19})\,(10^6)}{(1)\,(10^{-7})}} = 0.04\ \mathrm{deg.}$$

Thus, the scanning system will have a signal-to-noise power of unity for a temperature difference of 0.04 deg. In actual practice, we would wish to have much higher contrast, that is, a non-noisy image. A power signal-to-noise ratio of about 100/1 or 20 dB is usually adequate for a TV presentation, which means that a clear, high-contrast image would be obtained for temperature differentials 10 times the value of $NE\Delta T$, or 0.4 deg. Note that a factor-of-ten increase of ΔT will cause 100-fold increase of the output signal-to-noise power!

As a final note, we shall calculate the D^* of the detector in this application. To do this, we must assume a focal length, to fix the detector area. Here we shall use $f = 10$cm as a suitable value, in which case the detector area is Ωf^2 or 10^{-4} cm^2, that is, a detector of dimension approximately 0.1 mm. D^* then becomes

$$D^* = \frac{\sqrt{AB}}{(NEP)_{BL}} = \frac{\sqrt{AB}}{\sqrt{2h\nu B \cdot P_B}} = \frac{\sqrt{(10^{-4})\,(10^6)}}{\sqrt{2(0.124)\,(1.6 \times 10^{-19})(10^6)\,(10^{-7})}}$$

$$= 1.6 \times 10^{11}\ \mathrm{cm\ Hz}^{1/2} \cdot \mathrm{W}^{-1}$$

In the following paragraphs, we shall treat the general behavior of D^* for ideal background-limited photon detectors for comparison with this value, which was derived in an approximate manner.

2.7 D^* for an Ideal Detector

We now derive the value of D^* for an ideal detector in the presence of thermal radiation and as a function of the cutoff frequency or wavelength of the device. We consider the arrangement shown in Fig. 2.3, which shows an optical collection system exposed to an object field at temperature T. The size of the aperture is such that the full cone angle of received radiation at the detector is θ, and we shall take the detector area as A. We stipulate that the chamber that contains the detector is at a temperature much lower than that of the object field, so that the detected radiation is completely due to the power transmitted through the lens. We now introduce a new concept in our calculation, namely, that the radiation received by the detector throughout the cone angle θ is exactly the same as the radiation that would be received if a uniform radiating surface at temperature T replaced the back surface of the lens. This would be the case if the field of view of the system, Ω_{ext}, is completely encompassed by the object field. The proof is complicated for a generalized optical system, but in the case

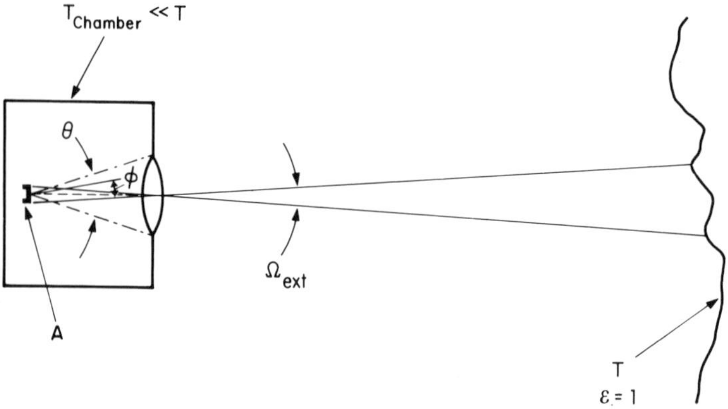

Fig. 2.3. Determination of thermal power at detector

of the simple thin lens, we merely note that the energy density u_ν at the outside surface of the lens within the angle Ω_{ext} is continous across the lens surface into the region contained within the detector cone of angle θ. Basically the lens takes all ray bundles within the small angle Ω_{ext} and bends them to direct them in the direction of the detector. Thus the detector "sees" radiation from the back of the lens that corresponds to the full irradiance that would be expected from a surface at temperature T. With this analytical approach, we may immediately write that the power striking the detector is

$$dP_{\nu,\Omega} = A \frac{2h\nu^3 d\nu}{c^2(e^{h\nu/kT} - 1)} \cos\phi \, d\Omega \,, \tag{2.20}$$

again from (1.11). Integrating over the cone angle θ yields

$$dP_\nu = A\,\frac{2h\nu^3 d\nu}{c^2(e^{h\nu/kT}-1)}\int_0^{\theta/2}\cos\phi\,(2\pi\sin\phi\,d\phi) = A\sin^2\frac{\theta}{2}\,\frac{2\pi h\nu^3 d\nu}{c^2(e^{h\nu/kT}-1)}\,. \quad (2.21)$$

We now digress and re-examine the terms in the expression for D^*, which was defined as

$$D^* = \frac{\sqrt{AB}}{NEP} = \frac{\sqrt{AB}}{\sqrt{2h\nu B\,P_B}} = \frac{\sqrt{A}}{\sqrt{2h\nu P_B}}\,.$$

We note, in this expression, that the NEP is a function of P_B; because we plan to integrate the effect of the radiation field over all frequencies greater than ν_c, we must note how P_B affects the noise output from the detector. Specifically, we rewrite the background power as

$$P_B = \bar{r}h\nu$$

because the rate of occurrence of photoevents determines the noise in the detector output. With this change, we write D^* as

$$D^* = \frac{\sqrt{A}}{\sqrt{2h\nu\,\bar{r}h\nu}} = \frac{\sqrt{A}}{h\nu\sqrt{2\bar{r}}}\,. \quad (2.22)$$

We return to the power expression (2.21) and obtain

$$d\bar{r}_\nu = A\sin^2\frac{\theta}{2}\cdot\frac{2\pi\nu^2 d\nu}{c^2(e^{h\nu/kT}-1)}\,, \quad (2.23)$$

which may be integrated to yield

$$\bar{r} = A\sin^2\frac{\theta}{2}\int_{\nu_c}^{\infty}\frac{2\pi\nu^2 d\nu}{c^2(e^{h\nu/kT}-1)} = A\sin^2\frac{\theta}{2}\cdot\frac{2\pi}{c^2}\cdot\left(\frac{kT}{h}\right)^3\int_{x_c}^{\infty}x^2 e^{-x}\,dx\,, \quad (2.24)$$

with $x = h\nu/kT \ll 1$.

Substituting this expression into (2.22) gives, finally,

$$D^* = \left\{\sin\left(\frac{\theta}{2}\right)\left[\frac{2\pi(kT)^5}{c^2h^3}\right]^{1/2}\cdot x_c(x_c^2+2x_c+2)^{1/2}\,e^{-x_c/2}\right\}^{-1}, \quad (2.25)$$

where the extra term in $h\nu$ has been incorporated into the dimensionless parameter x_c. A numerical evaluation of this expression yields

$$D^* = 1.3 \times 10^{11} \left(\frac{300}{T}\right)^{5/2} \frac{1}{\sin(\theta/2)} \cdot \frac{e^{x_c/2}}{x_c(x_c^2 + 2x_c + 2)^{1/2}}, \tag{2.26}$$

and a plot of the result is shown in Fig. 2.4, for $T = 300$ K. Note that the curve may be used for a different temperature by multiplying the ordinate by $(300/T)^{5/2}$ and the abscissa by $(300/T)$.

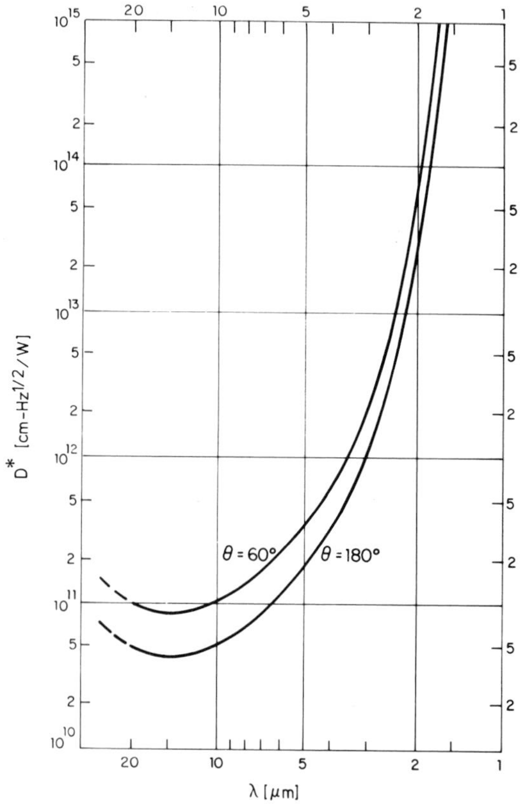

Fig. 2.4. D^* for ideal photon detector

Comparing the results of this rigorous calculation with the approximate value of D^* obtained in the previous section, we obtain

$$D^*\,(\theta = 53°) = D^*\,(180°)/0.45$$
$$= 1.2 \times 10^{11}\ \text{cm-Hz}^{1/2} \cdot \text{W}^{-1}$$

which is not far from the approximate value 1.6×10^{11} derived previously.

In conclusion, we should emphasize that the D^* shown in the figure is for a *perfect* detector operating into the specified acceptance angle at the specified wavelength. In later sections, we shall derive D^* obtainable for different types of detectors, realizing that the merit of a detector is measured by the ratio of its D^* to the value of D^* for a perfect detector. In particular, we shall find that many detectors have a constant D^*f^* product where f^* is the cutoff frequency of operation. Thus a detector may have ideal D^* for small bandwidths, but increasing the frequency response results in reduced detectivity.

Problems

2.1 Prove that $\overline{n^2} - \bar{n}^2 = \bar{n}$ for the Poisson distribution, $p(k,\bar{n}) = \bar{n}^k e^{-\bar{n}}/k!$ Note that

$$\bar{n} = \sum_{k=0}^{\infty} k\, p(k, \bar{n})$$

$$\overline{n^2} = \sum_{k=0}^{\infty} k^2 p(k,\bar{n})$$

and

$$e^{\bar{n}} = \sum_{k=0}^{\infty} \bar{n}^k/k!$$

[Hint: $(-1)! = \infty$ and $(-2)! = \infty$]

2.2 An infrared star at $T = 1000$K subtends an angle of 0.1 arc-s (1 arc-s = 5 microradians). Using a 1m diameter telescope, and a 10 μm detector with a 10% spectral bandwidth filter

a) Find the incident signal power on the detector.

b) If the atmospheric transmission is 80% in the 10 μm band, find the background power striking the detector, assuming that the detector field of view is 1 milliradian and that the atmosphere is at 300 K.

c) What is the maximum post-detection bandwidth which may be used to obtain a $(S/N)_P$ of 10 dB (10:1)?

2.3 A laser transmitter on a satellite at synchronous range, 40,000 km, transmits into a beam solid angle of 10^{-6} steradians. With a 1 m diameter receiver aperture and a perfect photon detector, what is the required average power transmitted for a bandwidth of 1000 MHz, a wavelength of 0.53 μm, and a $(S/N)_P$ of 15 dB.

2.4 Repeat the calculations of Section 2.6 but use a wavelength of 4 μm with the same fractional spectral bandwidth.

a) Find the background power.

b) Find $NE\Delta T$.

3. Coherent or Heterodyne Detection

In the previous chapter we discussed the detection of the incident power in an optical or infrared wave, basically ignoring the actual frequency of the wave except in the value of the quantum energy and the wavelength or frequency cutoff of the photon detector. We now consider heterodyne or coherent detection first demonstrated at optical frequencies by *Forrester* et al. (1955). By using a single-frequency local oscillator as a reference, we may convert the incoming wave's amplitude and phase to a low frequency in the rf or microwave region. In the following sections we treat the heterodyne case, first for simple constant-amplitude co-planar waves and then in a more general way for an arbitrary amplitude and phase distribution of the local oscillator over the detector surface. We then derive two important theorems that are applicable to heterodyne detection. These are the antenna theorem, which describes the angular response of the detector, and the mixing theorem, which allows the calculation of the mixing efficiency at any arbitrary surface.

3.1 Heterodyne Conversion and Noise

For our first simple derivation we assume a detector surface of area A having constant quantum efficiency η over the surface. Incident normal to this surface are two plane waves, whose E vectors lie in the plane of the surface and parallel to each other. We define the local oscillator field as $E_{\rm LO}$, and the signal field as $E_{\rm S}$, and then

$$E = E_{\rm LO} \cos \omega_{\rm LO} t + E_{\rm S} \cos \omega_{\rm S} t \,,$$

and we also know from the prior treatment that the photodetector current is given by

$$i(t) = \frac{e\eta P(t)}{h\nu} \,.$$

By use of the impedance of free space z_0, the power incident upon the surfaces may be written

$$P(t) = \frac{E^2(t)A}{z_0}$$

and the resultant detector current is

$$i(t) = \frac{e\eta A}{z_0 h\nu} (E_{LO} \cos \omega_{LO} t + E_S \cos_S \omega_S t)^2$$

$$= \frac{e\eta A}{z_0 h\nu} \left[\frac{1}{2} E_{LO}^2 (1 + \cos 2\omega_{LO} t) + \frac{1}{2} E_S^2 (1 + \cos 2\omega_S t) \right.$$

$$\left. + E_{LO} E_S \cos(\omega_S - \omega_{LO})t + E_{LO} E_S \cos (\omega_S + \omega_{LO})t \right],$$

where we have assumed that $(\omega_S - \omega_{LO})$ is much less than ω_{LO}, therefore justifying the use of a single value for hν. We note that the cosine terms in ω_{LO}, ω_S, and $(\omega_S + \omega_{LO})$ correspond to current variations at the optical frequency and can be ignored because the detector would not respond to such rapid variations. In actuality, a rigorous quantum-mechanical treatment (*Teich*, 1969) does not yield these terms, because the detector responds to the average power as measured over many cycles of the optical freqeuncy. Collecting the remaining terms and noting that

$$i_{LO} = \frac{e\eta A}{2z_0 h\nu} \cdot E_{LO}^2 ,$$

and

$$i_S = \frac{e\eta A}{2z_0 h\nu} \cdot E_S^2 ,$$

which are the dc currents due to the individual local oscillator and signal powers, we obtain finally

$$i(t) = i_{LO} + i_S + 2 \sqrt{i_{LO} i_S} \cdot \cos (\omega_S - \omega_{LO})t . \tag{3.1}$$

To calculate the $(S/N)_P$ at the output of the detector, we take the mean-square i. f. current at frequency $\omega_{i.f.} = (\omega_S - \omega_{LO})$ and obtain

$$\overline{i_{i.f.}^2} = (2 \sqrt{i_{LO} i_S})^2/2 = 2 i_{LO} i_S ,$$

where a factor of 1/2 has been introduced for the average of the square of the cosine. The noise current arises in the same manner as it did in the incoherent case, namely due to shot noise in the detection process; if $i_{LO} \gg i_{i.f.} \gg i_S$, as is generally the case, then

$$\overline{i_N^2} = 2\, e i_{LO} B ,$$

where B is the bandwidth of the i.f. channel following the detector. Taking the ratio of the signal and noise powers yields finally

$$\left(\frac{S}{N}\right)_P = \frac{\overline{i_{i.f.}^2}}{\overline{i_N^2}} = \frac{2 i_{LO} i_S}{2e\, i_{LO} B} = \frac{i_S}{eB} = \frac{\eta P_S}{h\nu B}, \tag{3.2}$$

where we have used the relationship $i_S = e\eta P_S / h\nu$.

Several important features of heterodyne detection may now be summarized. First, the i.f. signal power is directly proportional to the input optical power; second, the signal-to-noise ratio is inversely proportional to bandwidth, just as in rf and microwave systems; and third, the output i.f. spectrum for a broadband signal is an exact replica of the input optical spectral distribution, provided that the local oscillator is indeed a single-frequency wave with nonfluctuating phase. The last statement may be verified by replacing the constant-amplitude, constant frequency signal wave with a field amplitude and phase that are functions of time, always assuming of course that the spectral bandwidth of the signal is small compared to the center optical frequency.

The foregoing derivation is for the quite restricted case of constant amplitude and phase of the signal and local-oscillator fields over the detector surface. In addition, we assumed that both had the same polarization. We now formulate a much more general treatment, where we relax all of the stated restrictions, although, for simplicity, we shall assume that the wavefront associated with the signal or local-oscillator fields is parallel to the surface within a small angle θ such that $\cos\theta \simeq 1$. This requirement is necessary to equate the power density flow into the surface to the square of the electric field divided by the free-space impedance. Later, we shall show that this restriction may be relaxed when we prove the so-called mixing theorem. Using complex notation and spatial vector forms, we write the two fields as

$$\boldsymbol{E}_{\mathrm{LO}}(x,y,t) = \mathrm{Re}\,[\boldsymbol{E}_{\mathrm{LO}}(x,y)\,\mathrm{e}^{\mathrm{i}\omega_{\mathrm{LO}}t}]$$

$$\boldsymbol{E}_{\mathrm{S}}(x,y,t) = \mathrm{Re}\,[\boldsymbol{E}_{\mathrm{S}}(x,y)\,\mathrm{e}^{\mathrm{i}\omega_{\mathrm{S}}t}]$$

for a surface in the x,y plane. By direct comparison with our previous derivation, we may write that the incremental i.f. current produced in area dA is given by

$$di_{\mathrm{i.f.}} = \frac{e\eta(x,y)\,dA}{z_0 h\nu}\,\boldsymbol{E}_{\mathrm{LO}}\cdot\boldsymbol{E}_{\mathrm{S}}^{*}\,,$$

where we have also assumed that the quantum efficiency is a function of position on the detector surface. Performing the integration over the area and taking one-half the square of the magnitude yields the mean-square i.f. current,

$$\overline{i^2_{\mathrm{i.f.}}} = \frac{e^2}{2z_0^2h^2\nu^2}\left|\int \eta\boldsymbol{E}_{\mathrm{LO}}\cdot\boldsymbol{E}_{\mathrm{S}}^{*}\,dA\right|^2.$$

In a similar manner, we write the mean-square noise current as

$$\overline{i^2_{\mathrm{N}}} = 2ei_{\mathrm{LO}}B = \frac{e^2B}{z_0h\nu}\int \eta\boldsymbol{E}_{\mathrm{LO}}\cdot\boldsymbol{E}_{\mathrm{LO}}^{*}\,dA$$

and obtain a signal-to-noise ratio

$$\frac{\overline{i_{\text{i.f.}}^2}}{\overline{i_{\text{N}}^2}} = \frac{1}{2z_0 h\nu B} \frac{\left|\int \eta \boldsymbol{E}_{\text{LO}} \cdot \boldsymbol{E}_{\text{S}}\, dA\right|^2}{\int \eta \boldsymbol{E}_{\text{LO}} \cdot \boldsymbol{E}_{\text{LO}}^*\, dA} .$$

The signal power may be written as

$$P_{\text{S}} = \frac{1}{2z_0} \cdot \int \boldsymbol{E}_{\text{S}} \cdot \boldsymbol{E}_{\text{S}}^*\, dA .$$

Using this to eliminate z_0, we obtain

$$\left(\frac{S}{N}\right)_{\text{P}} = \frac{P_{\text{S}}}{h\nu B} \cdot \frac{\left|\int \eta \boldsymbol{E}_{\text{LO}} \cdot \boldsymbol{E}_{\text{S}}^*\, dA\right|^2}{\int \eta \boldsymbol{E}_{\text{LO}} \cdot \boldsymbol{E}_{\text{LO}}^*\, dA \int \boldsymbol{E}_{\text{S}} \cdot \boldsymbol{E}_{\text{S}}^*\, dA} . \tag{3.3}$$

Further manipulation leads to an important form of this expression,

$$\left(\frac{S}{N}\right)_{\text{P}} = \frac{P_{\text{S}}}{h\nu B} \cdot \frac{\left|\int \eta \boldsymbol{E}_{\text{LO}} \cdot \boldsymbol{E}_{\text{S}}^*\, dA\right|^2}{\int \eta \boldsymbol{E}_{\text{LO}} \cdot \eta \boldsymbol{E}_{\text{LO}}^*\, dA \int \boldsymbol{E}_{\text{S}} \cdot \boldsymbol{E}_{\text{S}}^*\, dA} \times \frac{\int \eta \boldsymbol{E}_{\text{LO}} \cdot \eta \boldsymbol{E}_{\text{LO}}^*\, dA}{\int \eta \boldsymbol{E}_{\text{LO}} \cdot \boldsymbol{E}_{\text{LO}}^*\, dA} .$$

We identify the two integral terms in the last expression as the product of the mixing efficiency m and the effective quantum efficiency η_{eff}. Thus, the signal-to-noise ratio is given by the product,

$$\left(\frac{S}{N}\right)_{\text{P}} = \frac{P_{\text{S}}}{h\nu B} \cdot m \cdot \eta_{\text{eff}} , \tag{3.4}$$

where

$$m = \frac{\left|\int \eta \boldsymbol{E}_{\text{LO}} \cdot \boldsymbol{E}_{\text{S}}^*\, dA\right|^2}{\int \eta \boldsymbol{E}_{\text{LO}} \cdot \eta \boldsymbol{E}_{\text{LO}}^*\, dA \int \boldsymbol{E}_{\text{S}} \cdot \boldsymbol{E}_{\text{S}}^*\, dA} , \tag{3.5}$$

and

$$\eta_{\text{eff}} = \frac{\int \eta \boldsymbol{E}_{\text{LO}} \cdot \eta \boldsymbol{E}_{\text{LO}}^*\, dA}{\int \eta \boldsymbol{E}_{\text{LO}} \cdot \boldsymbol{E}_{\text{LO}}^*\, dA} . \tag{3.6}$$

The last term is a property of the local-oscillator distribution, as modified by the distribution of η over the surface, and is thus independent of the signal field. The first term m, however, is a measure of the match between the incoming signal wave and the "effective" local-oscillator field. In fact, if we write the Schwarz inequality in the form

$$\left|\int f_1(x,y) f_2^*(x,y)\, dA\right|^2 \leq \int f_1(x,y) f_1^*(x,y)\, dA \int f_2(x,y) f_2^*(x,y)\, dA$$

and recognize that the equality holds only for

$$f_2(x,y) = k f_1(x,y)$$

with k a scalar constant, then we see that the mixing efficiency becomes unity if and only if

$$E_S(x,y) = k\eta(x,y)\, E_{LO}(x,y)\,.$$

Thus the mixing efficiency is always unity or less; the value is unity only when the signal field has a polarization, amplitude, and phase identical with the product of the quantum efficiency and the local-oscillator field at all points of the surface, at least within an arbitrary scalar multiplicative factor.

This concept of the mixing efficiency will be used in the next section to derive the antenna theorem for a heterodyne detector.

3.2 The Antenna Theorem

We now derive the antenna theroem, first postulated by *Siegman* (1966). In this treatment, we shall calculate the relative response of a detector to an incoming plane wave as a function of the angle of incidence with respect to the surface normal. We first derive the Fraunhofer diffraction integral, which gives the far-field radiation pattern produced by a specified electric-field distribution over a plane surface. By "far-field" we mean amplitude and phase of the electric field as a function of angle at a large distance from the surface radiation. Using the notation of Fig. 3.1, we define $E(x,y)$ as the complex electric field at the surface of the plane element, and for simplicity will assume linear polarization in the y direction. At a long distance from the surface, at a range of R, the incremental field contribution due to the surface electric field over area dA may be shown to be

$$dE_F(\theta_x,\theta_y) = \frac{\nu}{Rc}\, E(x,y)\, e^{-ikR} dA\,;\; k = \frac{2\pi\nu}{c}\,;\; \theta_x,\theta_y \ll 1\,,$$

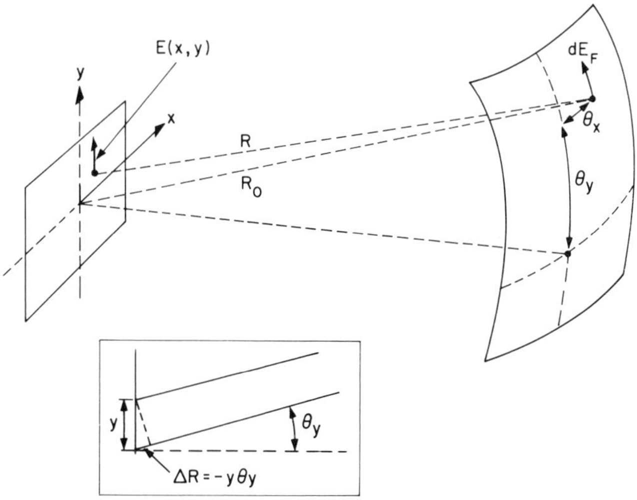

Fig. 3.1. Derivation of Fraunhofer integral

where the coefficient ν/Rc will be justified later. For calculation of the overall contribution of the surface field, we must obtain the variation of R as a function of x and y, because the phase delay associated with this R variation determines the complex addition of all the individual fields associated with each surface element dA. For the case in which θ_x and θ_y are small, the value of R may be shown to be

$$R = R_0 + \Delta R = R_0 - x\theta_x - y\theta_y$$

by the geometrical construction shown in the figure. This expression is valid provided that the change of θ_x or θ_y over the surface is negligible, that is, θ_x and θ_y are independent of position on the surface. It may be shown that this restriction is equivalent to stating that the error in ΔR is much less than a wavelength; simple algebra yields the requirement that $R_0 \gg d^2/\lambda$, where d is the maximum dimension of the radiating surface. This range is known as the far-field region and is the restriction under which the Fraunhofer integral is applicable. For closer ranges, the variation of phase has a quadratic term, and the appropriate integral is called the Fresnel integral. For small angles, the variables θ_x and θ_y may be identified with the x and y components of the propagation vector k,

$$\theta_x = \frac{k_x}{k}\ ;\ \theta_y = \frac{k_y}{k}$$

from which the phase term may be written,

$$-ikR = -ikR_0 + ik_x x + ik_y y\,.$$

Neglecting the overall phase shift ikR_0 and integrating over x and y yields finally the Fraunhofer integral

$$E_F(\theta_x,\theta_y) = E_F(k_x, k_y) = \frac{\nu}{Rc}\int E(x,y)\,e^{i(k_x x + k_y y)}\,dA\,. \tag{3.7}$$

This function is identical in form to the Fourier transform. We write the standard transform pair and the associated normalization property,

$$f(x,y) = \frac{1}{2\pi}\int\int F(k_x,k_y)\,e^{-i(k_x x+k_y y)}\,dk_x dk_y$$

$$F(k_x,k_y) = \frac{1}{2\pi}\int\int f(x,y)\,e^{+i(k_x x+k_y y)}\,dx dy$$

$$\int\int \left|f(x,y)\right|^2 dxdy = \int\int \left|F(k_x,k_y)\right|^2 dk_x dk_y\,.$$

From these relationships, we may immediately write the inverse transform

$$E(x,y) = \frac{Rc}{4\pi^2\nu}\int\int E(k_x,k_y)\,e^{-i(k_x x+k_y y)}\,dk_x dk_y\,. \tag{3.8}$$

Using the third equation of the Fourier-transform relationships, we obtain

$$\int\int \left|E(x,y)\right|^2 dxdy = \frac{R^2c^2}{4\pi^2\nu^2}\int\int E_F^2(k_x,k_y)\,dk_x dk_y\,. \tag{3.9}$$

To justify the coefficient ν/Rc, we use conservation of power to equate the far-field power flow to the flow outward across the radiating surface. Because the incremental area in the far field is given by

$$\begin{aligned} dA &= R^2 d\Omega = R^2 d\theta_x d\theta_y = R^2\,\frac{dk_x dk_y}{k^2} \\ &= \frac{R^2c^2}{4\pi^2\nu^2}\,dk_x dk_y\,, \end{aligned}$$

and the respective power flows are of the form

$$P = \frac{1}{2z_0}\int E^2 dA\,,$$

(3.9) indicates that power is indeed conserved, because the coefficient ν/Rc appears in the far-field expression.

We now prove the antenna theorem by noting that the mixing efficiency m is given by

$$m = \frac{\left| \int \eta \boldsymbol{E}_{\mathrm{LO}} \cdot \boldsymbol{E}_{\mathrm{S}}^{*} \, dA \right|^2}{\int \eta \boldsymbol{E}_{\mathrm{LO}} \cdot \eta \boldsymbol{E}_{\mathrm{LO}}^{*} \, dA \int \boldsymbol{E}_{\mathrm{S}} \cdot \boldsymbol{E}_{\mathrm{S}}^{*} \, dA} .$$

If we assume an incident plane wave of the form

$$\boldsymbol{E}_{\mathrm{S}} = \boldsymbol{E}_0 \, \mathrm{e}^{-\mathrm{i}(k_x x + k_y y)}$$

with $\boldsymbol{E}_0$ parallel to $\boldsymbol{E}_{\mathrm{LO}}$, where the negative sign in the exponent indicates an inward propagating wave, we obtain

$$m = \frac{\left| \int \eta \boldsymbol{E}_{\mathrm{LO}} \boldsymbol{E}_0 \, \mathrm{e}^{\mathrm{i}(k_x x + k_y y)} \, dA \right|^2}{\int |\eta \boldsymbol{E}_{\mathrm{LO}}|^2 \, da \int |\boldsymbol{E}_0|^2 \, dA} . \tag{3.10}$$

The numerator of this expression is simply the square of the magnitude of the Fraunhofer integral associated with a surface field of $\eta \boldsymbol{E}_{\mathrm{LO}}(x,y)$. In addition, our earlier derivation of the mixing efficiency indicates that the amplitude and phase of the i.f. current are directly proportional to the integral within the magnitude sign. Thus, the amplitude response of the detector is proportional to the Fraunhofer integral of the effective local-oscillator distribution. In conclusion then, the antenna theorem states that the complex angular response of the detector is given by the far-field pattern $\boldsymbol{E}_{\mathrm{F}}\,(\theta_x, \theta_y)$ that would be produced if we reverse time and let the impinging local-oscillator radiation propagate backward, into free space. Unfortunately, heterodyne detectors, per se, are seldom directly exposed to the incoming signal wave, an optical system usually being interposed between the two. In the next section, we shall show that the antenna theorem is valid not only at the detector surface but over any surface that completely encompasses the incoming local-oscillator and signal fields.

3.3 The Mixing Theorem

We now derive the mixing theorem, which states that the mixing efficiency m may be calculated over *any* surface that completely intercepts the local-oscillator and signal radiation that strikes the detector surface. We first note that for a local-oscillator distribution at the surface of the detector, there exists a local-oscillator field throughout space, corresponding to this surface distribution. We do not restrict this distribution to that in the far field but assume that there is a unique distribution given by $\boldsymbol{E}_{\mathrm{LO}}(x,y,z)$ throughout the medium, where we again use the complex vector field. The medium may also include lenses and reflectors, provided that they are lossless. In addition, we postulate that there is a set of orthonormal field functions $\boldsymbol{E}_{\mathrm{LO}k}(x,y,z)$ that correspond to a surface field $\boldsymbol{E}_{\mathrm{LO}k}(x,y)$ at the surface of the detector. Again we emphasize that our prior derivations assume that the signal and local-oscillator fields have wavefronts parallel to the surface; that is, that the power flow is normal to the surface. This requirement is necessary so that we may use the square field as a measure of the power flow. The mixing theorem originally derived by *Ross* (1970) uses the Poynting vector and thus avoids this restriction. We now specify that the functions $\boldsymbol{E}_{\mathrm{LO}k}$ are orthonormal by writing

$$\int \boldsymbol{E}_{\mathrm{LO}k} \cdot \boldsymbol{E}^*_{\mathrm{LO}m}\, dA = \delta_{km} \int \boldsymbol{E}_{\mathrm{LO}} \cdot \boldsymbol{E}^*_{\mathrm{LO}}\, dA \qquad \begin{array}{ll} \delta_{km} = 1 & k = m \\ \delta_{km} = 0 & k \neq m \end{array}$$

and we also expand the signal field at the surface,

$$\boldsymbol{E}_{\mathrm{S}}(x,y) = \sum_k a_{\mathrm{S}k}\, \boldsymbol{E}_{\mathrm{LO}k}(x,y)\,,$$

from which it follows that the signal field throughout the region is given by

$$\boldsymbol{E}_{\mathrm{S}}(x,y,z) = \sum_k a_{\mathrm{S}k}\, \boldsymbol{E}_{\mathrm{LO}k}(x,y,z)\,.$$

The mixing efficiency is given by

$$m = \frac{\left| \int \boldsymbol{E}_{\mathrm{LO}} \cdot \boldsymbol{E}^*_{\mathrm{S}}\, da \right|^2}{\int \boldsymbol{E}_{\mathrm{LO}} \cdot \boldsymbol{E}^*_{\mathrm{LO}}\, dA \int \boldsymbol{E}_{\mathrm{S}} \cdot \boldsymbol{E}^*_{\mathrm{S}}\, dA}\,,$$

where we have assumed that η is independent of x and y. We could include this surface variation by using an effective local-oscillator field $\eta \boldsymbol{E}_{\mathrm{LO}}$ expanded in an appropriate orthonormal set, but we shall avoid this complication for the sake of clarity. Using the properties of the local-oscillator set, we may rewrite the mixing efficiency as

$$m = \frac{|a_{\mathrm{SO}}|^2}{\sum_k |a_{\mathrm{S}k}|^2}\,,$$

which states that the effective signal power that results in i.f. output power is simply

$$P_{\mathrm{SO}} = |a_{\mathrm{SO}}|^2\, P_{\mathrm{LO}}\,,$$

whereas the total signal is

$$P_S = (\sum_k |a_{Sk}|^2)\, P_{LO}\,.$$

These last two relationships follow from the general expression

$$P_{LO} = \frac{1}{2z_0} \int \boldsymbol{E}_{LO} \cdot \boldsymbol{E}_{LO}^* \, dA\,.$$

To prove the mixing theorem, we construct an imaginary surface at a distance from the detector, which intercepts all of the impinging fields, and note that by conservation of energy, the power flow P_{SO} across the surface is independent of the position of the surface, provided that all of the signal field is intercepted by the surface. In fact, the power flow across *any* such surface consists of the independent terms P_{Sk}, each of which is continuous. This follows from the fact that any mode $\boldsymbol{E}_{LOk}$ transmits P_{LO} across any surface with conservation of energy, and that the a_{Sk} are all independent of the coordinates. Thus, the mixing theorem states that the mixing efficiency is independent of the surface over which it is evaluated *provided* that the surface *as well as* the detector intercepts all impinging radiation.

One unstated assumption was involved in this derivation, namely, that the signal and local oscillator fields are at the same frequency. This condition is inherent in our assumption that the sets of wave functions for the two fields are identical. Actually, a finite frequency difference will result in slightly different solutions of Maxwell's equations; *Ross* (1970) discusses the range of validity for a simple optical system. We shall consider a simple case by asking what is the range of validity for a warped surface adjacent to the detector surface. This derivation will also establish the required "flatness" of a detector in terms of its optimum detection capability.

In Fig. 3.2, we show an ideal flat detector surface that lies in the plane of the impinging radiation. We also note that the incoming local-oscillator wave and signal wave at the warped surface may be written as

$$\boldsymbol{E}'_{LO} = \boldsymbol{E}_{LO}\, e^{-ik_{LO}z}\ ; \quad \boldsymbol{E}'_S = \boldsymbol{E}_S\, e^{-ik_S z}\,,$$

where the $\boldsymbol{E}_{LO}$ and $\boldsymbol{E}_S$ are the values at the ideal flat surface. The mixing theorem states that for $k_S = k_{LO}$, m is invariant from one surface to the other. Taking into account the difference of k vectors leads to a new expression for the mixing term,

$$m \propto \int \boldsymbol{E}'_{LO} (\boldsymbol{E}'_S)^* \, dA = \int \boldsymbol{E}_{LO} \cdot \boldsymbol{E}_S^*\, e^{i(k_S - k_{LO})z}\, dA\,.$$

We conclude that the mixing is invariant only if

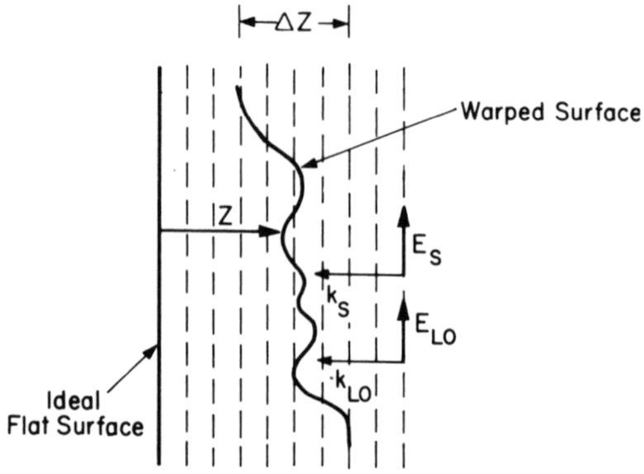

Fig. 3.2. Effect of finite frequency difference on mixing efficiency

$$(k_S - k_{LO})\,\Delta Z = \left(\frac{2\pi\nu_S}{c} - \frac{2\pi\nu_{LO}}{c}\right)\Delta Z \ll 1\,.$$

A little algebra then leads to the requirement,

$$2\pi\,\frac{\nu_S - \nu_{LO}}{c}\,\Delta Z = \frac{2\pi}{\lambda_{\mathrm{i.f.}}}\,\Delta Z \ll 1 \tag{3.11}$$

which simply states that the tilt or roughness of the surface must be small compared with the wavelength $\lambda_{\mathrm{i.f.}}$ corresponding to the i.f. frequency. The significant point is that the surface need *not* be optically flat, but only flat compared with the i.f. wavelength.

We conclude by combining the antenna theorem with the mixing theorem in the following example. In Fig. 3.3, we show a focussing optical system that is

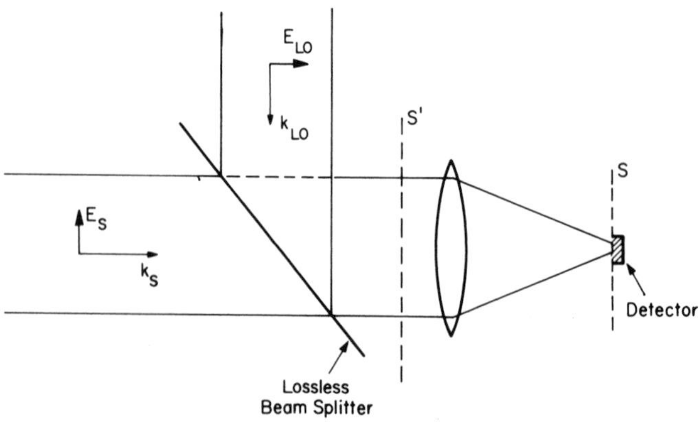

Fig. 3.3. Heterodyne detection system. Mixing efficiency may be calculated over S or S'

illuminated by a signal and a local oscillator, by use of a low-reflectance beam-splitting mirror such that the signal attenuation is small. Using the antenna theorem, we might calculate the far field of the local oscillator propagating backward out through the signal path from the detector. This would be quite complicated, but the mixing theorem tells us that we can propagate the local-oscillator wave backward from outside the focussing lens, *provided* that the detector intercepts all of the incoming radiation. Within this constraint, the receiver beam pattern is the Fraunhofer pattern associated with the local-oscillator field that leaves the beam splitter. If, for example, the field is constant, the far-field pattern or receiver-sensitivity pattern is simply the Airy function, that is, the far-field pattern for a constant-intensity circular aperture. The combination of the mixing theorem and antenna theorem thus has numerous applications in the design and analysis of heterodyne detection systems.

3.4 A Rederivation of Planck's Law

To conclude our discussion of heterodyne detection, we rederive Planck's law, using the modes associated with a given local-oscillator distribution, rather than the modes of a parallelopiped, as in Chapter 1. This derivation not only emphasizes the coupling of a heterodyne detector to thermal radiation, but also will be used as the basis for our analysis of thermal-radiation fluctuations in Chapter 8.

We start by asking what the effective ΩA product is for a detector with a specified local-oscillator distribution $\boldsymbol{E}_{\mathrm{LO}}(x,y)$ over the surface, that is, what is the average product of effective receiver area and beam solid angle. If we find this for an arbitrary mode, then the density of modes per unit area per unit solid angle will be obtained. We restrict the derivation to a distribution that has even symmetry and is maximum at the center of the detector, although the proof may be shown to be valid, with much more complicated mathematics, for any field distribution. With these restrictions in mind, we define the effective area of the heterodyne detector as the ratio of the i.f. *current* produced by a plane-parallel constant-amplitude signal wave to the *current per unit area* that would be produced if the local-oscillator field were constant and equal to its maximum value,

$$A_{\mathrm{eff}} = \frac{\int \boldsymbol{E}_{\mathrm{LO}}(x,y) \cdot \boldsymbol{E}_{\mathrm{S}}^{*}\, dA}{\boldsymbol{E}_{\mathrm{LO}}(0,0) \cdot \boldsymbol{E}_{\mathrm{S}}^{*}} = \frac{\int \boldsymbol{E}_{\mathrm{LO}}(x,y)\, dA}{\boldsymbol{E}_{\mathrm{LO}}(0,0)} \, ,$$

because $\boldsymbol{E}_{\mathrm{S}}$ is constant and its wavefront is parallel to the surface, that is, k_x and k_y are zero. We find the effective ΩA product by integrating the value of A_{eff} over the angle of incidence (θ_x, θ_y), taking into account the angular sensitivity,

$$f(\theta_x,\theta_y) = \frac{\boldsymbol{E}_{\mathrm{F}}(\theta_x,\theta_y)}{\boldsymbol{E}_{\mathrm{F}}(0,0)} \, ,$$

where, from (3.7) and (3.8), the functions $\boldsymbol{E}_{\mathrm{LO}}$ and $\boldsymbol{E}_{\mathrm{F}}$ may be written

$$\boldsymbol{E}_{\mathrm{LO}} = \frac{\nu R}{c} \int \boldsymbol{E}_{\mathrm{F}}(\theta_x,\theta_y)\, \mathrm{e}^{-\mathrm{i}(k_x x + k_y y)}\, d\Omega \text{ since } d\Omega = \left(\frac{c}{2\pi\nu}\right)^2 dk_x dk_y$$

$$\boldsymbol{E}_{\mathrm{F}} = \frac{\nu}{Rc} \int \boldsymbol{E}_{\mathrm{LO}}(x,y)\, \mathrm{e}^{+\mathrm{i}(k_x x + k_y y)}\, dA\,.$$

The area-angle product then becomes

$$(\Omega A)_{\mathrm{eff}} = \int f(\theta_x,\theta_y)\, A_{\mathrm{eff}}\, d\Omega = \frac{\int \boldsymbol{E}_{\mathrm{LO}}(x,y)\, dA \int \boldsymbol{E}_{\mathrm{F}}(\theta_x,\theta_y)\, d\Omega}{\boldsymbol{E}_{\mathrm{LO}}(0,0)\, \boldsymbol{E}_{\mathrm{F}}(0,0)}\,.$$

Substituting for $\boldsymbol{E}_{\mathrm{F}}$ at $(\theta_x,\theta_y) = (0,0)$ and $\boldsymbol{E}_{\mathrm{LO}}$ (0, 0) from the above expressions yields the final answer,

$$(\Omega A)_{\mathrm{eff}} = \frac{c^2}{\nu^2} = \lambda^2\,.$$

We now postulate that a complete set of orthogonal modes with $\boldsymbol{E}_{\mathrm{LO}}$ as the basis will completely fill the space to the right of the plane that contains the detector, and that the area-angle product of each of these modes is c^2/ν^2. Using the construction of Fig. 3.4 we construct a surface at radius R from the detector, where R is in the far field and the surface is completely reflecting. The allowed standing-wave modes as a function of frequency are then governed by the relation,

$$kR = \frac{2\pi\nu}{c} R = n\pi\,,$$

that is, there are an integral number of half wavelengths between the detector surface and the reflecting surface. The number of allowed standing-wave modes per unit frequency interval is then

$$\frac{dN}{d\nu} = \frac{2R}{c}\,.$$

Therefore the density of modes per unit frequency, solid angle, and area is

$$d^3N/d\nu\, d\Omega\, dA = (\nu^2/c^2) \times 2R/c\,.$$

The power flow in any mode consists of equal amounts P flowing outward and inward; the total energy in a mode is simply $2Pt$, where t is the time for the

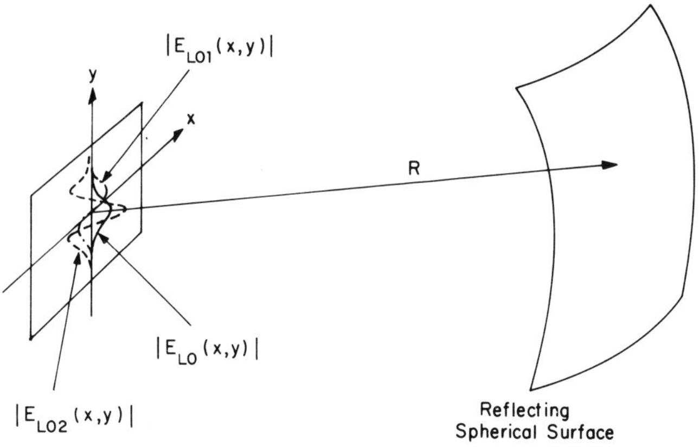

Fig. 3.4. Calculation of mode density in frequency space. Modes shown as an example are Gaussian-based modes of *Boyd* and *Gordon* (1961)

energy flow from the detector surface to the reflecting surface. Because $t = R/c$ and the energy per mode is given by the photon energy modified by the Bose-Einstein factor,

$$P = \frac{E}{2T} = \frac{c}{2R} \cdot \frac{h\nu}{(e^{h\nu/kT} - 1)},$$

the irradiance is

$$dI_{\nu,\Omega} = \frac{d^3N}{d\nu\, d\Omega\, dA} \cdot P\, d\nu = \frac{\nu^2}{c^2} \frac{h\nu\, d\nu}{(e^{h\nu/kT} - 1)},$$

which, except for a factor of two, corresponding to the two possible polarizations, is the same as the formula derived in Chapter 1. Thus, we see that the thermal-radiation field may be derived by use of any set of orthogonal field functions that completely fill space. In addition, we note that a heterodyne receiver couples into only *one* mode of space and thus to only one mode of the thermal-radiation field, whose power is given by the preceding formula.

Problems

3.1 A square detector of dimensions $d \times d$ has a local oscillator distribution E_{LO} of constant amplitude and phase.

a) Construct a set of orthogonal modes E_{LOm} of the form

$$E_{\mathrm{LOm}} = E_{\mathrm{LO}}\, \mathrm{e}^{\mathrm{i}(k_x x + k_y y)}$$

by finding the discrete values of k_x and k_y that satisfy the orthogonality condition.

b) Find the far-field distribution $E_{\mathrm{F}}(\theta_x,\theta_y)$ using Fraunhofer integral.

c) Show that the functions $E_{\mathrm{Fm}}(\theta_x,\theta_y)$ are orthogonal over a surface at radius R. Note: This proof requires contour integration using the residue theorem. It is tricky!

3.2 A heterodyne detection system operating at 10 μm wavelength observes the sun ($T = 5800$ K) with a receiver beam width smaller than the sun subtense. What is the $(S/N)_{\mathrm{P}}$ for this system, assuming the radiation is spectrally filtered so that the solar signal power is much less than the local oscillator power?

3.3 In the sketch below, show that the mixing efficiency is zero. How would you solve this dilemma by appropriate combination of the two detector currents?

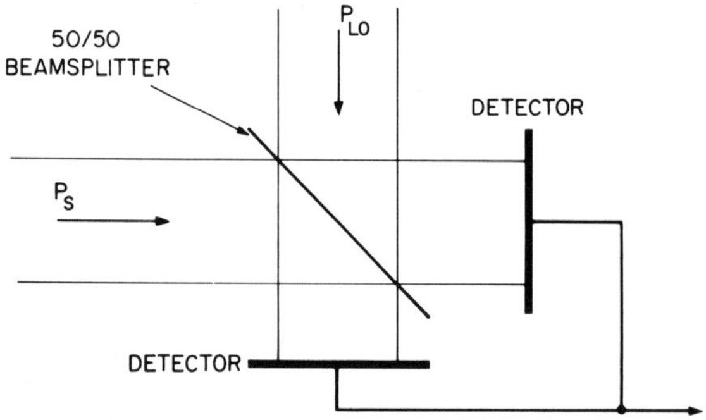

3.4 A heterodyne detector is illuminated with local oscillator power P_{LO}. Find the conversion gain $P_{\mathrm{i.f.}}/P_{\mathrm{S}}$ in terms of the optical frequency and detector load resistance R. Assume unity quantum and mixing efficiency. What is the conversion gain for $P_{\mathrm{LO}} = 10$ mW, $\lambda = 1$ μm, and $R = 50\ \Omega$?

4. Amplifier Noise and Its Effect on Detector Performance

Until now we have calculated signal-to-noise ratios assuming that the detector is followed by a noiseless amplifier. In actual use, however, we must take into account amplifier noise which is characterized by the noise figure of the particular amplifier chosen. Considering the circuit of Fig. 4.1, we define the noise figure as

$$F = \left(\frac{S}{N}\right)_{\mathrm{IN}} \Big/ \left(\frac{S}{N}\right)_{\mathrm{OUT}},$$

where S and N represent the electrical signal and noise powers with the source resistor $R_{\mathrm{D}}(T)$ at a temperature of 300 K. The gain G of the amplifier takes into account any mismatch between the load R_{D} and the amplifier input impedance. In fact, the noise figure of any amplifier is a function of R_{D} and is often a minimum value for R_{D} not equal to the input impedance, that is, under mismatched conditions. Taking into account the fact that G includes the loss due to mismatch, and defining P_{exc} as the excess noise added by the amplifier, we find

$$F = \frac{S_{\mathrm{IN}}}{S_{\mathrm{OUT}}} \times \frac{N_{\mathrm{OUT}}}{N_{\mathrm{IN}}} = \frac{1}{G} \times \frac{(G\,kT_{300}B + P_{\mathrm{exc}})}{kT_{300}B} = 1 + \frac{P_{\mathrm{exc}}}{G\,kT_{300}B}$$

because the *available* power from the 300 K resistor is $kT_{300}B$. When the amplifier is operated with a real detector load at temperature T_{R}, the effective input noise becomes

$$(N_{\mathrm{IN}})_{\mathrm{EFF}} = \frac{GkT_{\mathrm{R}}B + P_{\mathrm{exc}}}{G} = [kT_{\mathrm{R}} + (F - 1)kT_{300}]\,B$$

and the output signal-to-noise ratio is

$$\left(\frac{S}{N}\right)_{\mathrm{OUT}} = \frac{S_{\mathrm{IN}}}{N_{\mathrm{D}} + (N_{\mathrm{IN}})_{\mathrm{EFF}}},$$

where S_{IN} and N_{D} are the available powers produced by the detector signal current and noise current. These are given by

$$S_{\mathrm{IN}} = \overline{i_{\mathrm{S}}^2}\, R/4$$
$$N_{\mathrm{D}} = \overline{i_{\mathrm{N}}^2}\, R/4$$

and the final result is

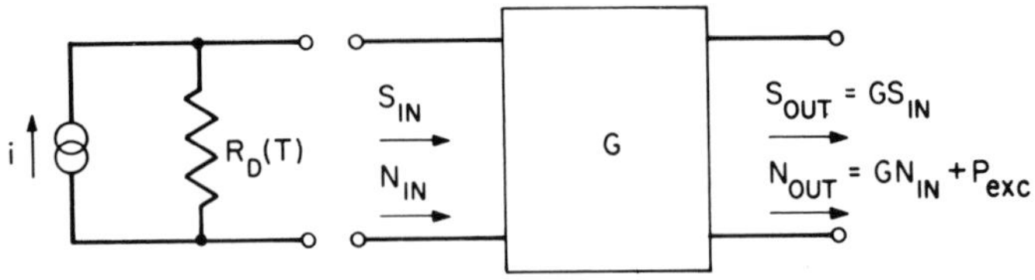

Fig. 4.1. Detector-amplifier circuit

$$\left(\frac{S}{N}\right)_{\mathrm{OUT}} = \frac{\overline{i_S^2}}{\overline{i_N^2} + 4kT_N B/R}\,; \quad T_N = T_R + (F-1)\,T_{300}\,. \tag{4.1}$$

This key equation will be used throughout our further discussions of amplifier noise and its effect on detector performance.

4.1 Amplifier Noise in Incoherent Detection

Using the above expression, we may now calculate the signal-to-noise power for an incoherent detector, using

$$\overline{i_S^2} = \frac{e^2\eta^2 P_S^2}{(h\nu)^2}$$

$$\overline{i_N^2} = \frac{2\eta e^2(P_S + P_B)B}{h\nu}\,.$$

The result, after some manipulation, becomes

$$\left(\frac{S}{N}\right)_P = \frac{\eta P_S^2}{2h\nu B(P_S + P_B)\left[1 + \left(\frac{2h\nu}{e}\right)\left(\frac{kT_N}{e}\right)\left(\frac{1}{\eta R}\right)\left(\frac{1}{P_S + P_B}\right)\right]}\,. \tag{4.2}$$

The crossover point between ideal detection and amplifier-noise-limited detection occurs when

$$(P_S + P_B) = \left(\frac{2h\nu}{e}\right)\left(\frac{kT_N}{e}\right)\left(\frac{1}{\eta R}\right), \tag{4.3}$$

whereas for amplifier noise dominating the system, the result is

$$\left(\frac{S}{N}\right)_P = \frac{e^2\eta^2 P_S^2}{(h\nu)^2\,(4kT_N B/R)}\,.$$

Thus the noise-equivalent power for the latter case becomes

$$(NEP)_{\mathrm{A.L.}} = \frac{2h\nu}{e\eta}\sqrt{\frac{kT_{\mathrm{N}}B}{R}}\,. \tag{4.4}$$

We note that this value of NEP is again proportional to the square root of the bandwidth. In addition, as we shall see later, the allowable terminating resistance R is inversely proportional to C, the detector capacitance, which, in turn, is usually proportional to area. Thus the NEP is again proportional to $(AB)^{1/2}$, as in the background-limited case.

4.2 Amplifier Noise in Coherent or Heterodyne Detection

For the coherent-detection case we use $i_{\mathrm{i.f.}}$ as the signal current and write the signal-to-noise expression as

$$\left(\frac{S}{N}\right)_{\mathrm{P}} = \frac{\overline{i_{\mathrm{i.f.}}^2}}{\overline{i_{\mathrm{N}}^2} + 4kT_{\mathrm{N}}B/R}$$

with, as before,

$$T_{\mathrm{N}} = T_{\mathrm{R}} + (F-1)T_{300}\,.$$

From Chapter 3, the appropriate mean square currents are given by

$$\overline{i_{\mathrm{i.f.}}^2} = 2i_{\mathrm{LO}}i_{\mathrm{S}}$$
$$\overline{i_{\mathrm{N}}^2} = 2ei_{\mathrm{LO}}B$$

and the resultant $(S/N)_{\mathrm{P}}$ is

$$\left(\frac{S}{N}\right)_{\mathrm{P}} = \frac{2i_{\mathrm{LO}}i_{\mathrm{S}}}{2ei_{\mathrm{LO}}B + 4kT_{\mathrm{N}}B/R} = \frac{i_{\mathrm{S}}}{eB(1 + 2kT_{\mathrm{N}}/ei_{\mathrm{LO}}R)}\,.$$

Using $i_{\mathrm{S}} = e\eta P_{\mathrm{S}}/h\nu$, we obtain finally

$$\left(\frac{S}{N}\right)_{\mathrm{P}} = \frac{\eta P_{\mathrm{S}}}{h\nu B(1 + 2kT_{\mathrm{N}}/ei_{\mathrm{LO}}R)}\,;$$

with $i_{\mathrm{LO}} = e\eta P_{\mathrm{LO}}/h\nu$, the result is

$$\left(\frac{S}{N}\right)_{\mathrm{P}} = \frac{\eta P_{\mathrm{S}}}{h\nu B(1 + 2kT_{\mathrm{N}}h\nu/e^2\eta P_{\mathrm{LO}}R)}\,, \tag{4.5}$$

where we have assumed unit mixing efficiency and $\eta = \eta_{eff}$ as given by (3.5). Thus, we obtain detector noise-limited or ideal heterodyne detection if

$$P_{LO} \gg \left(\frac{2kT_N}{e}\right)\left(\frac{h\nu}{e}\right)\left(\frac{1}{\eta R}\right).$$

Here we note the exact equivalence with the incoherent case (4.3) where the background quantity $(P_S + P_B)$ must satisfy the same requirement for true background-limited detection. The physical significance is that the noise due to the total power falling on the detector dominates the excess noise generated in the following amplifier.

As an example of the requirement dictated by this expression, consider a wideband heterodyne detector (several hundred MHz bandwidth) terminated by a 50 Ω load and operating at a wavelength of 10 μm. The local-oscallator requirement then becomes

$$P_{LO} \gg \left(\frac{2kT_N}{e}\right)\left(\frac{h\nu}{e}\right)\left(\frac{1}{\eta R}\right) = 2(.026)\,(0.124) \cdot \frac{1}{50\eta} = \frac{0.12}{\eta}\ \mathrm{mW}\,,$$

where we assume $T_N = 300$ K, a feasible number for a 3 dB (2:1) noise figure and a cooled detector load. Thus, even for this broad bandwidth requirement, the required local-oscillator power is only of the order of several milliwatts. For a wavelength of 1 μm, the required power is increased by a factor of ten.

Problems

4.1 A *perfect* photon detector has an output capacitance of 10 pF and is to be used for detection of a 100 MHz bandwidth signal. Using a receiver amplifier with a noise figure of 5 dB (3/1) and a room-temperature detector load, what local oscillator power will just make the detector noise equal to the amplifier noise. Use a wavelength of 10 μm.

4.2 A 1 μm wavelength detector, with $\eta = 1$ has an amplifier with a noise figure of 0.4 dB and is terminated in a liquid-nitrogen (77 K)-cooled load of 1 MΩ.

a) What is the value of T_N?
b) Find $(NEP)_{AL}$ for a bandwidth of 10 kHz.
c) At what signal power does the detector become signal-noise limited.

4.3 For the same operating conditions as in Problem 4.2 use heterodyne detection and find

a) The minimum detectable power.
b) The required local oscillator power.

5. Vacuum Photodetectors

In the visible and near infrared regions of the spectrum, vacuum photodetectors are the most commonly used photon detectors, either as simple photodiodes or with electron multiplication as in photomultipliers. The efficiency of the photoelectric effect unfortunately diminishes beyond 1 µm wavelength where in the most recent photocathodes quantum efficiencies of 1 to 2 percent have been obtained under very special conditions. As the photon energy increases, the available quantum efficiency rises to as high as 20 to 30% from 0.5 µm into the ultraviolet. Beyond 1 µm wavelength, the choice of detector is limited to semiconductor devices, if the ultimate in sensitivity is to be approached.

5.1 Vacuum Photodiodes

For analysis of a vacuum photodiode, we consider a parallel-plate structure consisiting of a photocathode and positively biased anode, with applied voltage V. A practical, device might have a special configuration, but the important factor is the time of flight between any point on the cathode surface and the anode. For optimum frequency response, this transit time should be minimized and constant for all points on the cathode. Otherwise the pulse response of the device will be distorted. Returning to the plane-parallel case, we shall calculate the current transform $i(\omega)$ for use in a calculation of the shot-noise frequency spectrum and note that this response function is also the amplitude and phase response of the detector vs frequency, because it is the current response to an input impulse.

Referring to Fig. 5.1, we note that the current pulse at the anode is triangular because the instantaneous current during the electron passage is proportional to velocity and thus to the acceleration times the time. (We neglect any initial emission velocity, which is at most thermal). Because the area under the $i(t)$ curve must equal the charge e, the current is given by

$$i = \frac{2e}{\tau}\left(\frac{t}{\tau}\right) = \frac{2e}{\tau^2} \cdot t\,,$$

where τ is the transit time. Because the acceleration is $a = (e/m)E$, the transit time is obtained as follows:

$$d = \frac{1}{2}a\tau^2 = \frac{1}{2}\frac{e}{m}\frac{V}{d}\tau^2$$

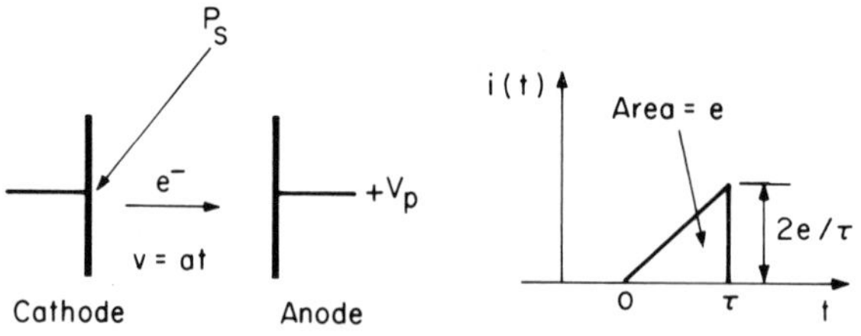

Fig. 5.1. Current inpulse in a vacuum photodiode

$$\tau = d\sqrt{\frac{2}{(e/m)V}},$$

where d is the cathode-anode spacing. Solving now for $i(\omega)$ we obtain

$$i(\omega) = \int_0^{\tau} i(t)\,\mathrm{e}^{-\mathrm{i}\omega t}\,dt = \frac{2e}{(\omega\tau)^2}\,[(1 + \mathrm{i}\omega\tau)\,\mathrm{e}^{-\mathrm{i}\omega\tau} - 1]\;;$$

taking the square of the magnitude yields

$$|i(\omega)|^2 = \frac{4e^2}{(\omega\tau)^4}\left[4\sin^2\left(\frac{\omega\tau}{2}\right) + (\omega\tau)^2 - 2\omega\tau\sin\omega\tau\right] \triangleq e^2F(\omega\tau)$$

which is the normalized power response of the detector vs frequency. To calculate the noise current, we write

$$i_N^2 = 2\bar{r}B\,|i(\omega)|^2 = 2\frac{i}{e}B\,|i(\omega)|^2 = 2eiB[F(\omega\tau)]$$

from (2.6). Here the bandwidth B is assumed to be a narrow segment taken at a given value of frequency or, for low frequencies, B is the full bandwidth over the region where the quantity in brackets is unity. This latter condition holds for $\omega\tau \ll 1$. The response curve derived above may be found in *Davenport* and *Root* (1958, p. 123). The half-power point occurs very close to the value $\omega\tau = \pi$, so that the effective cutoff frequency is given by

$$f_c = \frac{\omega_c}{2\pi} = \frac{1}{2\tau}.$$

Thus far, the vacuum photodiode behaves as an ideal detector, except for the limited quantum efficiency and finite frequency cutoff. We now investigate extraneous noise currents and also the performance when coupled to an amplifier. First, the extra noise current is supplied by thermionic emission from the photocathode as governed by the Richardson-Dushman equation, which relates the emission current to the work function and the cathode temperature. Strictly speaking, the work function and photoelectric threshold energy are equal for only a simple metal; newer semiconductor photocathodes behave differently. However, the simple calculation that follows gives an order-of-magnitude approximation for the expected effects. The "dark" current, that is, the current in the absence of illumination, is given by

$$I_{\mathrm{d}} = 120 T^2 \, \mathrm{e}^{-h\nu_{\mathrm{c}}/kT} \ \mathrm{A\,cm^{-2}},$$

where ν_c is the cutoff frequency. This current, which is again a random sequence of electrons, produces shot noise equivalent to a background power given by

$$(P_{\mathrm{B}})_{\mathrm{eff}} = \frac{h\nu I_{\mathrm{d}}}{e\eta}.$$

We may then write the NEP for the device as

$$NEP = \sqrt{\frac{2h\nu B P_{\mathrm{B}}}{\eta}} = \frac{h\nu}{\eta e}\sqrt{2eI_{\mathrm{d}}B}.$$

As a typical example, we take a photodetector with 1 µm cutoff wavelength and a quantum efficiency of 1%. The dark current for a 1 cm^2 cathode area, from the dark-current equation, is 3×10^{-15} A cm^{-2} at $T = 300$ K, and the *NEP* becomes

$$NEP = \frac{(1.24)}{\eta}\sqrt{2.(1.6 \times 10^{-19})(3 \times 10^{-15})B} = \frac{4 \times 10^{-17}}{\eta}\sqrt{B}\ \mathrm{W}$$

with a D^* therefore of

$$D^* = \frac{\sqrt{AB}}{NEP} = (2.5 \times 10^{16})\,\eta = 2.5 \times 10^{14}.$$

This sensitivity can be improved without limit by cooling the photocathode, but we shall see shortly that amplifier noise imposes severe limits on the actual performance.

Taking into account amplifier noise and using the relation between the dark current and the effective background power, we may write the overall expression for $(S/N)_{\mathrm{P}}$ for the vacuum photodiode as

$$\left(\frac{S}{N}\right)_{\mathrm{P}} = \frac{\eta P_{\mathrm{S}}^2}{2h\nu B\left(P_{\mathrm{S}} + P_{\mathrm{B}} + \dfrac{h\nu I_{\mathrm{d}}}{\eta e}\right)\left[1 + \left(\dfrac{2h\nu}{e}\right)\left(\dfrac{kT_{\mathrm{N}}}{e}\right)\left(\dfrac{1}{\eta R}\right)\left(\dfrac{1}{P_{\mathrm{S}} + P_{\mathrm{B}} + \dfrac{h\nu I_{\mathrm{d}}}{\eta e}}\right)\right]}, \tag{5.1}$$

using (4.1). Also, for negligible dark current and background power, the amplifier-noise-limited *NEP* becomes

$$(NEP)_{\mathrm{AL}} = \frac{2h\nu}{\eta e}\sqrt{\frac{kT_{\mathrm{N}}B}{R}}, \tag{5.2}$$

as in our earlier treatment of the ideal photodetector. We now consider a typical diode structure and calculate the appropriate value of this quantity, showing in the process that the photodiode is generally amplifier-noise limited except for extremely high background or signal powers.

Let us assume a 1 cm^2 cathode and anode area, spaced at 1 cm, with an applied voltage of 300 V. Then the transit time τ is given by

$$\tau = d\sqrt{\frac{2}{\left(\frac{e}{m}\right)V}} = 10^{-2}\sqrt{\frac{2}{(1.76\times 10^{11})(300)}} = 2\times 10^{-9}\,\mathrm{s},$$

corresponding to a frequency cutoff f_c of $1/2\tau$ or 250 MHz. Now if we are to obtain the full bandwidth capability, the load resistance must be small enough so that the diode shunt capacitance does not limit the electrical response of the diode output circuit. Calculating the capacitance as

$$C = \frac{\varepsilon_0 A}{d} = \frac{(8.85\times 10^{-12})(10^{-4})}{10^{-2}} = 8.85\times 10^{-14}\ \mathrm{F},$$

we find that the load resistance is given by

$$R = \frac{1}{2\pi f_c C} = 7200\ \Omega,$$

assuming of course that there is no extraneous coupling capacitance associated with the amplifier connection. Assuming now a 1 μm quantum efficiency of 1 %, and an excess noise T_N of 300 K, we obtain for the *NEP*,

$$(NEP)_{\mathrm{AL}} = \frac{2h\nu}{\eta e}\sqrt{\frac{kT_{\mathrm{N}}B}{R}} = 2\times 10^{-10}\ \mathrm{WHz}^{-1/2},$$

which is seen to be five orders of magnitude less than that calculated for the dark-current-limited case. Of course, we can improve the operation by decreasing the required frequency response, thus increasing *R*, but the diode would still not approach the dark-current value unless the load resistance were increased by *ten* orders of magnitude. Another comparison may be made by calculating the D^* for this device,

$$D^* = \frac{\sqrt{AB}}{NEP} = \frac{1}{2 \times 10^{-10}} = 5 \times 10^9 \ \mathrm{cm} \cdot \mathrm{Hz}^{1/2} \cdot \mathrm{W}^{-1}$$

A comparison of this value with the ideal D^* shown in Fig. 2.3 again indicates that the detectivity is many orders of magnitude less than the extrapolated value that might be obtained at the wavelength of 1 μm. As a consequence, vacuum photodiodes are limited to large-signal applications when operated in the incoherent mode. But note that heterodyne or coherent operation is perfectly feasible, based upon the calculations in Section 4.2.

5.2 Photomultipliers

The solution to the amplifier-noise problem is to employ current gain in the photodetector, prior to amplification in a standard amplifier. This is accomplished by a set of "dynodes" or secondary-emitting electrodes, which multiply the incident current striking the dynode surface. A schematic diagram of such a structure is shown in Fig. 5.2, where the curved electrode structure is designed to minimize dispersion in the transit times of the individual electrons as they traverse the structure. The noise current associated with i_0 in the diagram is exactly the same as was calculated for the photodiode. Successive amplification of the current in the following stages multiplies both the signal and noise currents by the overall power gain of the structure, which for electron multiplication of m per stage gives a power gain of m^{2N}. If m were a constant, the signal and noise powers would be amplified equally and the output signal-to-noise ratio would be the same as that of the photodiode, but the effect of the subsequent amplifier would be diminished by the reciprocal of the photomultiplier gain. Actually, the multiplication process is slightly noisy; we treat this noise in the following paragraph (*van der Ziel*, 1954).

We first define the probability of k secondaries being emitted from the dynode as $P(k)$ and then define the mean and mean-square value of the multipli-

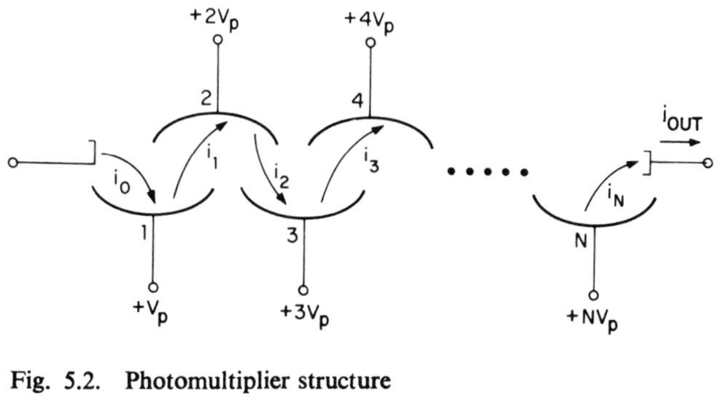

Fig. 5.2. Photomultiplier structure

cation factor by

$$\bar{k} = m = \sum_{k=0}^{\infty} kP(k) \; ; \; \overline{k^2} = \sum_{k=0}^{\infty} k^2 P(k) \, .$$

We then ask for the mean-square noise current that leaves the first stage due to the incident current i_0. If the dynode had a constant multiplication k, then the mean-square fluctuation in i_1 would be

$$\overline{i_{N1}^2} = 2(ke)(ki_0)B = 2k^2 \; ei_0 B \, ,$$

because the current consists of a random set of pulses, each containing charge ke, and with total current ki_0. But k is not a constant, so we must average over the distribution in k,

$$\overline{i_{N1}^2} = \sum_{k=0}^{\infty} 2k^2 P(k) ei_0 B = 2 \; ei_0 B \, \overline{k^2} \, .$$

The mean-square noise current at the output of the device due to fluctuations of i_0 is given by that fluctuation multiplied by the full power gain m^{2N}. The *excess* noise introduced by the current from the first stage i_1 is given by

$$(\overline{i_{N1}^2})_{\text{excess}} = 2ei_0 B(\overline{k^2} - m^2) \, ,$$

where we note that a constant value of k would result in zero excess noise. Multiplying by the remaining gain, which affects i_1, we obtain

$$(\overline{i_{N1}^2})_{\text{OUT}} = m^{(2N-2)} \cdot 2ei_0 B \, (\overline{k^2} - m^2) \, .$$

Proceeding to the second stage, we note that the input current $i_1 = mi_0$, so that the contribution from the current i_2 is

$$(\overline{i_{N2}^2})_{\text{OUT}} = m^{(2N-4)} \cdot 2 \; emi_0 B \, (\overline{k^2} - m^2) \, ,$$

and

$$(\overline{i_{N3}^2})_{\text{OUT}} = m^{(2N-6)} \cdot 2 \; em^2 i_0 B \, (\overline{k^2} - m^2) \, .$$

Proceeding through the stages up to the Nth electrode, we obtain

$$(\overline{i_N^2})_{\text{OUT}} = 2ei_0 B \left[m^{2N} + (\overline{k^2} - \overline{m^2}) \sum_{1}^{N} m^{2N-2n} \, m^{n-1} \right]$$

$$= 2ei_0 B\, m^{2N} \left[1 + \overline{(k^2} - m^2) \sum_1^N m^{-n-1} \right].$$

Using the following manipulations,

$$\sum_1^N m^{-n-1} = \frac{1}{m} \sum_1^N m^{-n} = \frac{1}{m} \left[\left(\sum_0^N m^{-n} \right) - 1 \right]$$

$$= \frac{1}{m} \left(\frac{1}{1 - \frac{1}{m}} - 1 \right) = \frac{1}{m(m-1)} .$$

because

$$\sum_0^N x^{-n} = \frac{1}{1 - x^{-1}} \quad \text{for} \quad N \gg 1 ,$$

we obtain finally

$$\overline{(i_N^2)}_{OUT} = 2ei_0 B \cdot m^{2N} \left[1 + \frac{\overline{(k^2} - m^2)}{m(m-1)} \right] = 2ei_0 B \cdot m^{2N} \left[\frac{\overline{k^2} - m}{m(m-1)} \right] .$$

We now define the noise factor Γ as

$$\Gamma = \left[\frac{\overline{k^2} - m}{m(m-1)} \right] \tag{5.3}$$

and note that the power gain of the device is $G = m^{2N}$, whereas the mean-square output noise current is

$$\overline{(i_N^2)}_{OUT} = 2ei_0 B \Gamma G . \tag{5.4}$$

For multiplication factors up to the order of 5, the probability distribution $P(k)$ is found experimentally to be close to Poisson and thus we may write, from (2.2),

$$\overline{k^2} - m^2 = m ,$$

which leads to

$$\Gamma = \frac{\overline{k^2} - m}{m(m-1)} = \frac{m^2}{m(m-1)} = \frac{m}{m-1} .$$

Thus, for a multiplication factor $m = 5$, the noise factor is only 1.25. For higher values of m, the distribution has a greater mean-square fluctuation, but the m in the denominator balances out this effect to such an extent that noise factors are at most 1.5 for typical photomultipliers. For a general discussion of photomultipliers see *Melchior* et al., (1970) and cited references.

We now write the S/N formula for the photomultiplier, using (4.1). First the signal current becomes

$$\overline{i_S^2} = Ge^2\eta^2 P_S^2/(h\nu)^2$$

and the noise current is

$$i_N^2 = 2\eta\Gamma Ge^2\left[P_S + P_B + \left(\frac{h\nu I_d}{\eta e}\right)B\right].$$

The result is

$$\left(\frac{S}{N}\right)_P = \frac{\eta P_S^2}{2\Gamma h\nu B\left(P_S+P_B+\frac{h\nu I_d}{\eta e}\right)\left[1+\left(\frac{1}{\Gamma G}\right)\left(\frac{2h\nu}{e}\right)\left(\frac{kT_N}{e}\right)\left(\frac{1}{\eta R}\right)\left(\frac{1}{P_S+P_B+\frac{h\nu I_d}{\eta e}}\right)\right]}. \tag{5.5}$$

We note that the crossover point for amplifier-limited detection now occurs at a much lower value of effective background or signal power, specifically

$$\left(P_S + P_B + \frac{h\nu I_d}{\eta e}\right) = \left(\frac{1}{\Gamma G}\right)\left(\frac{2h\nu}{e}\right)\left(\frac{kT_N}{e}\right)\left(\frac{1}{\eta R}\right). \tag{5.6}$$

Because G may be as high as 10^7, photomultipliers are typically signal, background, or dark-current limited.

Problems

5.1 A photomultiplier has a gain G, dark current I_D, quantum efficiency η, and noise factor Γ. Derive an expression for the signal-to-noise power ratio including the effects of background power and amplifier noise, assuming a load resistance R and noise temperature T_N.

5.2 What is the signal-noise-limited *NEP*. That is, neglect I_D and the amplifier noise.

5.3 What is the background-limited *NEP*?

5.4 What is the dark-current-limited *NEP*?

5.5 What is the amplifier-noise-limited *NEP*?

5.6 A photomultiplier with a cutoff wavelength of 1 μm is cooled to reduce the dark-current-limited *NEP*. At what temperature will $(NEP)_{DL}$ be reduced by 20 dB (100/1) from the $T = 300$ K value?

6. Noise and Efficiency of Semiconductor Devices

We now treat the two basic semiconductor photodetectors—the photoconductor and the photodiode. In the latter case, we also discuss the avalanche photodiode, which is the semiconductor analog of the photomultiplier. The photoconductor is a device composed of a single uniform semiconductor material; the incident optical power is measured by monitoring the conductance. The change of conductance is produced by creation of free carriers in the material by the presence of the radiation. In contrast, the photodiode is a *p-n* junction, in which photoinduced carriers in the vicinity of the junction modify the current-voltage characteristic so that the radiation level may be measured. In the case of the reverse-biased semiconductor photodiode, the behavior is almost identical to that of the vacuum photodiode.

6.1 Photoconductors

There are two basic types of photoconduction, extrinsic and intrinsic. In the extrinsic case, the photoconduction is produced by absorption of a photon at impurity levels and consequent creation of a free electron in the case of an *n*-type photoconductor, or a free hole in the case of a *p*-type photoconductor. The cutoff wavelength in either case is determined by the appropriate ionization energy for the impurity, as indicated schematically in Fig. 6.1. The extrinsic case is characterized by a low absorption coefficient α, because of the small number of available impurity levels, where α is defined by the attenuation equation,

$$P(x) = P(0)\,\mathrm{e}^{-\alpha x}\,.$$

Typical values of α for an impurity or extrinsic photoconductor are of the order of 1 to 10 cm^{-1}.

In contrast, the intrinsic photoconductor depends upon an across-the-gap transition that creates a free electron and hole simultaneously; the cutoff frequency is determined by the bandgap energy $\mathscr{E}_\mathrm{G}$. In this case, the absorption coefficient is much larger, due to the large number of available electron states associated with the valence and conduction band. Here α is of the order of 10^4 cm^{-1} for photoexcitation near the bandgap energy.

We initially treat the induced photocurrent and noise subject to a very important restriction, namely, that the temperature of the device is low enough so that the photoinduced carriers are the *only* carriers available for conduction. This requirement is satisfied if T_D (the detector temperature) $\ll h\nu_\mathrm{c}/k$, because

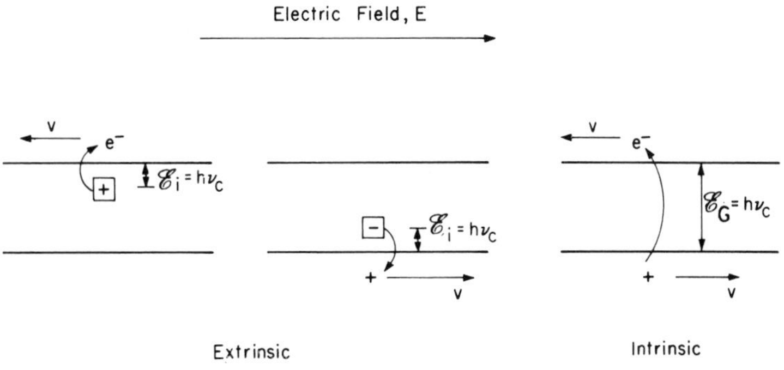

Fig. 6.1. Photoconduction process

thermal excitation of the carriers depends upon the same ionization energy as the photoexcitation process. This restriction is essential for optimum sensitivity and allows a much simpler treatment of the appropriate noise parameters.

Starting with the structure shown in Fig. 6.2, we assume uniform illumination of the detector with power P and wish to determine the current i in the presence of an applied voltage V. For this calculation, we shall assume an n-type extrinsic device, the p-type device behavior being identical. Later, we shall generalize the

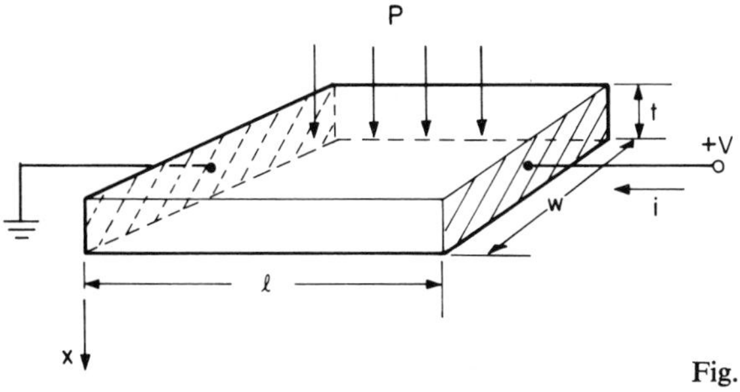

Fig. 6.2. Photoconductor structure

results to the intrinsic case. The current flowing between the two contacts decreases exponentially in the x direction because of the absorption of the incident radiation. The total current is given by

$$i = \int_0^t jdA = \int_0^t (nev)\,(wdx) = wev \int_0^t n(x)\,dx\,,$$

where $n(x)$ is the electron density associated with the radiation field $P(x)$. If the electrons have a mean lifetime τ, then the rate of recombination is given by $n(x)/\tau$, and the generation rate is given by

$$r = \left(\frac{1}{h\nu}\right) \frac{\alpha P(x)}{wl} = \frac{n(x)}{\tau} ;$$

the last equality comes from the requirement that the system is in steady state. Note also that the term $\alpha P/wl$ is simply the power absorbed per unit volume. Taking into account any reflection of the incident power by the surface reflectance r, we write

$$P(x) = P(1 - r)\, \mathrm{e}^{-\alpha x}$$

and solving for $n(x)$, we obtain

$$n(x) = \frac{\alpha P(1 - r)\tau}{wl \cdot h\nu}\, \mathrm{e}^{-\alpha x} .$$

The velocity of the electrons is $\mu_n E$, where μ_n is the electron mobility; substituting into the current equation, we obtain

$$i = wev \int_0^l n(x)\, dx = \frac{\alpha P(1 - r)\tau e \mu_n E}{lh\nu} \int_0^l e^{-\alpha x}\, dx .$$

Defining the quantum efficiency η as

$$\eta = (1 - r)\alpha \int_0^l \mathrm{e}^{-\alpha x}\, dx$$

leads to the final result,

$$i = \frac{\eta e P}{h\nu} \left(\frac{\mu_n E \tau}{l}\right) = \frac{\eta e P}{h\nu} \left(\frac{v\tau}{l}\right) .$$

We also define the quantity in parentheses as the photoconductive gain g. This quantity, when examined physically, is the ratio of the mean carrier lifetime τ to the transit time l/v between the two contacts. This is a measure of the effective charge transported through the external circuit per photoinduced electron. It may have a value greater than unity if the lifetime is greater than the transit time. What happens physically is that the excited electron *leaves* the semiconductor at the positive contact and by charge neutrality a replacement electron *enters* at the opposite contact. On the average, however, any particular electron "event" has a lifetime τ.

Actually, the recombination process is probabilistic and is governed by the expression,

$$p(t) = \frac{1}{\tau}\, \mathrm{e}^{-t/\tau} ,$$

where $p(t)$ is the probability that a carrier will recombine after a time t. Note that, if we calculate the average recombination time τ, we obtain

$$\tau = \int_0^\infty tp(t)\,dt = \int_0^\infty \frac{t}{\tau}\,e^{-t/\tau}\,dt = \tau \int_0^\infty xe^{-x}\,dx = \tau\,,$$

as would be expected from the definition. We now use this recombination probability expression to calculate the frequency response of the device. If we apply a spike impulse of radiation, the resultant current will be of the form

$$i = i_0 e^{-t/\tau}\,,$$

and the response in the frequency domain will be

$$i(\omega) = \int_0^\infty i_0 e^{-t/\tau}\,e^{-i\omega t}\,dt = \frac{i_0\tau}{1 + i\omega\tau}\,,$$

leading to a power response vs frequency of

$$|\,i(\omega)\,|^2 = \frac{i_0^2\tau^2}{1 + \omega^2\tau^2}\,.$$

The cutoff frequency f_c is then determined by setting $\omega\tau = 1$, or the power response falls to one-half at,

$$f_c = \frac{1}{2\pi\tau}\,.$$

An important observation may now be made, that the product of the gain g and the cutoff frequency f_c, is a constant. Thus, the gain may be increased only at the expense of a lower cutoff frequency, if τ is varied while the length and carrier velocity remain fixed.

We now consider the noise processes in such a device, in particular the noise associated with the discrete electron generation and recombination events. Although photoconductors are burdened with many types of noise, such as contact noise, surface noise, etc., we shall ignore these extra effects because they generally occur at low frequencies and do not radically limit device performance, except below about 1 kHz in certain types of applications. We should emphasize, however, that the photoconductor is a resistor in the true thermal sense and exhibits the standard Johnson or thermal noise. Although we shall not prove this fact here, we point out that the electrons (or holes) are in thermal equilibrium with the lattice and therefore with the temperature bath of the device, provided that the *collision* time is much shorter than the lifetime or transit time. This condition is almost universally fulfilled, so that even though the carrier concentration in the presence of a radiation field is much greater than the normal thermally generated value, the effective temperature of the electrons is still given by T_D. Thus, the device behaves as a resistor with resistance V/i, and generates thermal noise in the characteristic manner. What we shall now consider is the *current* noise produced in addition to the Johnson noise in the presence of an applied voltage. We may think of this noise as the *fluctuation* of R, which causes a fluctuation of i with a fixed voltage V.

We derive this noise current by noting that the current from the device consists of a set of random pulses, whose arrival times, are Poisson distributed, *but* whose pulsewidths also fluctuate because of the probabilistic behavior of the recombination process. In contrast to the simple photodiode, each pulse does *not* contain charge e but contains average charge ge. This may be understood by realizing that $g = 1$ corresponds to a typical electron making a full transit of the device thus supplying one electron to the external circuit. The individual pulsewidths are distributed according to the probability distribution for recombination; the contribution to the noise current from pulses having width t is

$$d\,(\overline{i_N^2}) = 2\,ge\left(\frac{t}{\tau}\right)\,diB\,, \tag{6.1}$$

where g is the average gain. The current element di, associated with a pulse of width t is simply

$$di = \left(\frac{t}{\tau}\right)p(t)\;idt = i\left(\frac{t}{\tau}\right)\cdot\frac{1}{\tau}\,\mathrm{e}^{-t/\tau}\,dt\,,$$

because the contribution to the current is weighted by the pulse width, t. This leads to the total mean-square noise current, after integration over all times t,

$$\overline{i_N^2} = 2geiB\int_0^\infty \frac{t^2}{\tau^3}\,\mathrm{e}^{-t/\tau}\,dt$$
$$= 2geiB\int_0^\infty x^2\mathrm{e}^{-x}\,dx = 4\,geiB\,, \tag{6.2}$$

where we have used the equality

$$\int_0^\infty x^n\,\mathrm{e}^{-x}\,dx = n!\,.$$

Thus, the noise current behaves in a manner similar to that of the photodiode, except that the effective charge is now ge rather than e, and the fluctuation of effective charge introduces a factor of two into the total fluctuation. This type of noise is called *g-r* noise, for generation-recombination; it is also applicable for the current noise in a semiconductor whose carriers are thermally excited rather than optically excited. Unfortunately, a rigorous treatment of g-r noise in semiconductors involves the severe complication of several independent generation-recombination processes, each of which involves recombination centers and impurity levels in the bandgap. We shall restrict our treatment to the simple case of one recombination time, either to impurity levels, in the extrinsic case, or directly across the gap in the intrinsic case.

We now determine the frequency spectrum of the noise power, realizing that we would expect a spectral shape the same as the power response previously derived. Unfortunately, our derivation in Chapter 2, using the spectrum of the individual pulse, is no longer applicable because the individual pulses have random width. To derive the frequency spectrum, we must therefore average the individual-pulse response spectra over the weighted contribution as a function of pulsewidth t. We do this by modifying (6.1) to include the spectral power density associated with the individual pulsewidth t. Thus the noise spectral density becomes

$$\overline{i_N^2}(f) = 2\,gei \int_0^\infty \frac{t^2}{\tau^3}\,e^{-t/\tau}\,\frac{\sin^2(\omega t/2)}{(\omega t/2)^2}\,dt\,,$$

which is the same as (6.2), except for the inclusion of the normalized spectral power response to a square pulse. The latter term is obtained by taking the square of the magnitude of the amplitude response function of (2.8). The integration may be performed by writing the integral as

$$\int_0^\infty \frac{4}{(\omega\tau)^2}\,e^{-t/\tau}\sin^2\left(\frac{\omega t}{2}\right)\frac{dt}{\tau} = \frac{2}{(\omega\tau)^2}\int_0^\infty e^{-x}\,[1-\cos(\omega\tau x)]\,dx\,,$$

yielding

$$\frac{2}{(\omega\tau)^2}\left[1-\frac{1}{1+(\omega\tau)^2}\right] = 2\left[\frac{1}{1+(\omega\tau)^2}\right].$$

The final answer, as we expected, is

$$\overline{i_N^2}(f) = 4\,gei\left(\frac{1}{1+\omega^2\tau^2}\right).$$

We now present an alternative derivation of the noise in a semiconductor in the presence of an applied voltage, by treating the device as a resistor with fluctuating resistance. We again consider the structure of Fig. 6.2 and calculate the fluctuation of resistance as determined by the fluctuation of N, the total number of free carriers present in the device. Now N is given by

$$N = \frac{\eta P}{h\nu}\,\tau$$

because this is the total rate of production of carriers multiplied by the mean carrier lifetime. The average carrier density is $\bar{n} = N/wtl$; writing the conductivity of the material as $\sigma = \bar{n}e\mu$, we obtain the resistance,

$$R = \frac{l}{\sigma A} = \frac{l}{\bar{n}elwt} = \frac{l^2}{eN\mu}\,.$$

We now wish to calculate the fluctuation of current through the device in the presence of an applied voltage V, because this is a measure of the resistance fluctuation. The current is simply

$$i = \frac{V}{R} = \frac{eN\mu V}{l^2}$$

and its mean-square fluctuation is given by

$$\overline{i_N^2} = \overline{i^2} - \bar{i}^2 = \frac{e^2\mu^2 v^2}{l^4}\,(\overline{N^2} - \bar{N}^2)\,.$$

The fluctuation of N is that of a large classical ensemble of particles, because we are assuming that the carrier density is small enough so that the distribution is Maxwell-Boltzmann; that is, the state occupancy in the free carrier band is much less than unity. This assumption was inherent in our use of a mobility applicable to each individual carrier. The mean-square fluctuation in such a case is just the mean value of N, so that we may write

$$\overline{i_N^2} = \frac{e^2\mu^2 V^2}{l^4}\,N = \frac{e\mu V}{l^2}\,i\,,$$

where we have used the expression for i to eliminate N. This quantity is the total fluctuation of the current. We wish to find its spectral density $\overline{i_N^2}\,(f)$, so that we can relate it to the preceding derivation. We do this by realizing that the fluctuation of N has a characteristic time constant τ associated with the recombination process and that the spectrum is thus of the same form as that derived earlier for the signal response and the noise response. Therefore

$$\overline{i_N^2} = \int \overline{i_N^2}(f)\,df = \int \overline{i_N^2}(0)\left(\frac{1}{1+\omega^2\tau^2}\right)df\,,$$

and the integral over the response function is found to be $1/4\tau$. The final result

$$\overline{i_N^2}(0) = 4\tau\overline{i_N^2} = \frac{4ei\mu V\tau}{l^2} = 4\,gei\,,$$

where we have used the relationship $g = \mu E\tau/l = \mu V\tau/l^2$, which defines the photoconductive current gain. Here we have derived the noise in terms of the fluctuation of the *number* of carriers in the device and the resultant fluctuating resistance. A similar derivation may be used to treat the Johnson noise, present even at zero bias, but in terms of the fluctuation of the *velocity* distribution of the carriers. In this case, the characteristic time is not the recombination time, but the collision time, because this is the characteristic time for a random change of velocity. A derivation of this noise term may be found in *Yariv* (1971, p. 258).

We conclude our discussion of photoconductor noise with two rather important generalizations based upon the foregoing treatments. These are:

1) The current noise, or *g-r* noise, is given by (6.2), using the total current, even if part of the current arises from the normal thermally excited carriers when the temperature is not low enough to preclude thermal excitation. This statement is valid if and only if there is one principal generation process with associated time constant τ, that is, a single impurity-conduction (or valence) band transition for the extrinsic semiconductor or an across-the-gap transition for the intrinsic case. Otherwise, the fluctuations of the carrier density will be governed by additional time constants, associated for example with recombination or trapping centers within the gap. Because the foregoing treatment indicates a proportionality of the noise current spectral density to τ in these processes, allowed transitions to other levels with time constants much smaller than the lifetime have negligible effect upon the noise and may be ignored. The principal argument we make is that the thermally generated carriers recombine with the same characteristic time as those generated optically, so that our treatment need make no distinction in terms of their effect upon the noise current.

2) The Johnson or thermal noise associated with the optically excited carriers is the same as that expected for a normal conductor, if and only if the collision time is much less than either the lifetime or the transit time through the device. Although not proven, the *Yariv* treatment referenced above requires only that the carrier distribution be a normal thermal distribution with defineable temperature T. Because the carriers thermalize in a collision time, the velocity distribution is normal thermal unless they recombine or leave the detector in a comparable period. With rare exceptions, these restrictions are satisfied by common photoconductive materials, so that we shall use these two rules in the following treatment.

We now derive the general expression for the signal-to-noise ratio for a photoconductor in the presence of amplifier noise, with the following definitions. R is the parallel combination of the detector and load resistance, at effective temperature T_R, which is used in the definition of T_N. I_D is the "dark" current, or current that flows in the device due to thermally excited carriers. Then the appropriate signal and noise currents may be written as

$$\overline{i_S^2} = \eta^2 g^2 e^2 P_S^2/(h\nu)^2$$

$$\overline{i_N^2} = 4\,geiB = \frac{4\eta g^2 e^2 B}{h\nu}\left(P_S + P_B + \frac{h\nu I_D}{\eta g e}\right),$$

and using (4.1), we obtain

$$\left(\frac{S}{N}\right)_P = \frac{\eta P_S^2}{4h\nu B\left(P_S + P_B + \dfrac{h\nu I_D}{\eta g e}\right)\left[1 + \left(\dfrac{h\nu}{e}\right)\left(\dfrac{kT_N}{e}\right)\left(\dfrac{1}{\eta g^2 R}\right)\left(\dfrac{1}{P_S+P_B+\dfrac{h\nu I_d}{\eta g e}}\right)\right]}. \tag{6.3}$$

Setting $(S/N)_P = 1$, we then find the *NEP* for the several cases; signal-noise limited, background-noise limited, dark-current limited, and amplifier-noise limited. These are

$$(NEP)_{SL} = \frac{4h\nu B}{\eta}, \tag{6.4}$$

$$(NEP)_{BL} = 2\sqrt{\frac{h\nu B P_B}{\eta}}, \tag{6.5}$$

$$(NEP)_{DL} = \frac{h\nu}{\eta e}\sqrt{\frac{2eI_D B}{g}}, \tag{6.6}$$

$$(NEP)_{AL} = \frac{2}{g}\frac{h\nu}{\eta e}\sqrt{\frac{kT_N B}{R}}. \tag{6.7}$$

We now examine in detail the behavior of a typical detector in terms of the different operating regimes. First, we note that the detector is signal, background or dark-current limited, that is, *not* amplifier limited if

$$\left(P_S + P_B + \frac{h\nu I_D}{\eta g e}\right) \gg \left(\frac{h\nu}{e}\right)\left(\frac{kT_N}{e}\right)\left(\frac{1}{\eta g^2 R}\right).$$

In the special case when we let the detector act as its own load resistance, that is, with R equal to the detector resistance V/i, solving for i gives a value of

$$\frac{1}{R} = \frac{1}{R_D} = \frac{i}{V} = \frac{1}{V}\frac{\eta g e}{h\nu}\left(P_S + P_B + \frac{h\nu I_D}{\eta g e}\right).$$

Substitution into the previous equation then yields the requirement that

$$V \gg \frac{kT_N}{ge},$$

in which case amplifier noise does not limit the performance and (6.4), (6.5), or (6.6) is applicable. Unfortunately, the typical value of R obtained in this mode is extremely high for available detector materials, as we shall show in the following treatment.

Let us first consider the signal-noise limited case, which is the ultimate in detector sensitivity. In this case, we again use the nonamplifier-limited-noise criterion, but neglect background and dark current,

$$P_S \gg \left(\frac{h\nu}{e}\right)\left(\frac{kT_N}{e}\right)\left(\frac{1}{\eta g^2 R}\right).$$

For signal-limited detection, $P_S = 4h\nu B/\eta$; substitution into the previous equation yields the requirement,

$$R \gg \left(\frac{kT_N}{e}\right)\frac{1}{eg^2 B},$$

where R includes a shunt load resistance R_L. Using the expression for the gain g, this may be written as

$$R \gg \left(\frac{kT_N}{e}\right)\frac{l^2}{4e\mu^2 E^2 \tau^2 B}.$$

We now consider a typical photoconductor, such as copper-doped germanium operating at 20 K or lower, with the material parameters

$$\begin{aligned} \mu &= 10^4 \mathrm{cm}^2\ \mathrm{V}^{-1}\ \mathrm{s}^{-1} \\ \tau &= 10^{-6}\ \mathrm{s} \\ l &= 0.1 \mathrm{mm} \\ E_{max} &= 100\ \mathrm{V\ cm}^{-1}, \end{aligned}$$

where the maximum electric field is limited by impact ionization of the impurity states due to the energy gained between collisions. We also assume, for simplicity, an amplifier-noise temperature T_N of 300 K. The resultant resistance required for signal-noise-limited detection becomes

$$R \gg \frac{(0.026)\,(10^{-4})}{4(1.6 \times 10^{-19})\,(10^8)\,(10^4)\,(10^{-12})\,B} = \frac{4 \times 10^{12}}{B}\,\Omega,$$

which is inconveniently great, considering the ever-present capacitance of the detector and amplifier input. Although the detector response extends to the order of 1 MHz because of the short lifetime, even a narrow-tuned network somewhere in the response band would be unrealizable because of the capacitative loading. Thus, at least with the parameters assumed above, signal-limited detection is not feasible with photoconductors. Unfortunately, the assumed parameters are typical of practically all available materials, at least in terms of the product $\mu E_{max}\tau$. Diminishing the length l would reduce the resistance, but the fabrication would be extremely difficult and l must always be greater than several wavelengths.

Obviously, as the background increases, the detector eventually becomes background limited and should obey the curves given in Fig. 2.3 (reduced by a factor of $1/\sqrt{2}$ because of the extra *g-r* noise), with appropriate angular field of view, wavelength, and background temperature. There is, however, a maximum value of D^*, determined by the detector-material characteristics and the amplifier noise. This "ceiling" may be calculated as follows. We again consider a square detector of dimension l and assume a thickness t, given by $1/\alpha$. The quantum efficiency is then approximately unity, actually $(1-1/e) = 63\%$, and the capacitance of the device is $\kappa\varepsilon_0/\alpha$ where ε_0 is the permittivity expressed in F cm^{-1}, $\varepsilon_0 = 8.85 \times 10^{-14}$. Writing the amplifier noise-limited NEP as

$$NEP = \frac{2h\nu}{g\eta e}\sqrt{\frac{kT_N B}{R}},$$

and writing the resistance as limited by the cutoff frequency and the capacitance as

$$R = \frac{1}{2\pi f_c C} = \frac{\alpha}{2\pi\kappa\varepsilon_0 f_c},$$

we obtain

$$NEP = \frac{2h\nu l}{\mu E\tau e}\sqrt{\frac{2\pi\kappa\varepsilon_0 k T_N B f_c}{\alpha}}.$$

The value of D^* then becomes

$$D^* = \frac{\sqrt{AB}}{NEP} = \frac{l\sqrt{B}}{NEP} = \frac{\mu E\tau}{2(h\nu/e)}\sqrt{\frac{\alpha}{2\pi\kappa\varepsilon_0 k T_N f_c}};$$

after substitution of the same parameters as used in the signal-limited case, assuming operation at 10 µm, the final result is

$$D^* = 2.7 \times 10^{16}\ \sqrt{1/f_c},$$

where we have assumed a value for α of 1 cm^{-1} and $\kappa = 10$. The general formula, taking into account different values of μ, E, τ, λ, T_N, α, and κ, is

$$D^* = 2.7 \times 10^{16} \frac{\mu}{10^4} \frac{E}{10^2} \frac{\tau}{10^{-6}} \sqrt{\frac{\alpha}{(\kappa/10)\,(T_N/300) f_c}} \, .$$

It may be seen that for longer wavelengths, D^* as limited by amplifier noise is well above the values shown in Fig. 2.3, at least for a cutoff frequency commensurate with the lifetime. At the shorter wavelengths or with low backgrounds and fields of view, the amplifier noise becomes the limiting factor in detector performance.

Up to this point, we have considered the extrinsic, or impurity-doped photoconductor. The intrinsic case is exactly the same, except for three qualifications. First, we replace the gain by an expression that includes the contribution of both carriers,

$$g = \frac{(\mu_n \tau_n + \mu_p \tau_p)\, E}{l} \, .$$

In general, the mobility of one type of carrier dominates, so that the expression usually reduces to the same form as that for the extrinsic case. Second, the absorption coefficient α is much greater than that for the extrinsic case, and we must derive the amplifier-limited D^* in terms of the stray capacitance of the detector and amplifier leads. It should be noted that the capacitance for a 1 cm^{-1} absorption is 0.09 pF; thus, larger values of α would yield capacitances much less than are attainable in any realizable circuit. Thus, the derived D^* is just about the maximum attainable, at least in terms of any increase of the absorption coefficient. A third factor in the intrinsic case is the common operation of intrinsic detectors with finite dark resistance. That is, the temperature required to "freeze-out" the free carriers is often too low to be practical in many applications. In this case (and the equations also apply to the extrinsic case), we determine the dark-current limited D^* in terms of the known or measured dark resistance. A typical example of this case is HgCdTe for 10 μm detection at liquid-nitrogen temperature (77 K). Considering again a square detector, we note that the dark resistance R_D is independent of dimension for a given material and temperature. We then write the dark-current-limited *NEP* as

$$(NEP)_{DL} = \frac{h\nu}{\eta e} \sqrt{\frac{2eI_D B}{g}} = \frac{h\nu}{\eta e} \sqrt{\frac{2eVB}{gR_D}} \, ,$$

assuming, of course, that there is only one dominant recombination process with lifetime τ. Using the expression for the gain, $g = \mu V \tau / l^2$, we obtain

$$(NEP)_{DL} = \frac{h\nu}{\eta e} \sqrt{\frac{2eBl^2}{R_D \mu \tau}} \, ;$$

solving for D^* yields

$$D^* = \frac{\eta e}{h\nu}\sqrt{\frac{R_D \mu\tau}{2e}}\,.$$

We again see the importance of the product $\mu\tau$, which, for an example, we shall again take as 10^{-2} ($\mu = 10^4$, $\tau = 10^{-6}$). For unit quantum efficiency and $\lambda = 10$ μm, the result is

$$D^* = 1.4 \times 10^9 \sqrt{R_D}\,.$$

Thus if the detector has a dark resistance of 100 Ω, D^* becomes 1.4×10^{10}, a value approximately the same as the background-limited case for a full 180 degree field of view at $T = 300$ K. At the shorter wavelengths, D^* decreases with decreasing wavelength because of the photon-energy term; however, the dark resistance increases dramatically because of the lower carrier concentration at the broader energy gap. Thus the dark-current-limited behavior follows closely the shape of the background-limited D^* curve for a given ambient temperature of the detector.

6.2 Semiconductor Photodiodes

A semiconductor *p-n* junction diode may also be used as a photodetector and behaves in many ways like a vacuum photodiode. The device may be operated in the reverse-biased mode or in the so-called "photovoltaic" mode although, in the literature, photovoltaic sometimes refers to either mode of operation. We shall reserve the term photovoltaic for the nonbiased diode, in which the open-circuit voltage is a measure of the signal flux. In contrast, in the reverse-biased case, the light-induced reverse current is the measure of the incident flux, just as in the vacuum photodiode. We start by reviewing *p-n* junction theory, and deriving expressions for the junction capacitance and the voltage-current relationship.

Consider the energy-band diagram of Fig. 6.3, which shows a *p-n* junction at zero applied voltage with metallic contacts at each end. There are a large number of free electrons in the *n*-type material that are neutralized by positive donor impurities N_D; similarly, there are a large number of holes in the *p*-type material that are neutralized by negative acceptor impurities N_A. The dashed line is the Fermi level, which is a measure of the energy, $\mathscr{E}_F$, for which the probability of state occupancy is one-half, as defined in (1.2). In equilibrium, the Fermi energy must be constant throughout the semiconductor and the metallic contacts. The difference of electrostatic potential between the *p*- and *n*-type material is just enough to maintain charge neutrality in the medium. As drawn in the figure, the donor states are empty of electrons, because they lie above the Fermi level; but because the conduction band contains many more available states, there is till

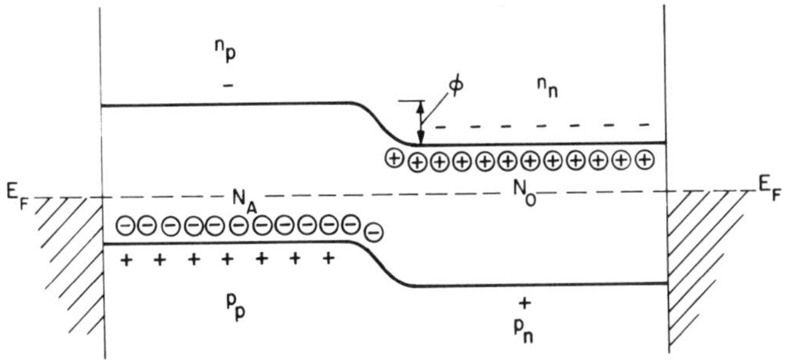

Fig. 6.3. Energy diagram for *p-n* junction

a large electron density, even for the low value of the occupancy probability. The Fermi level is thus clamped in the vicinity of the donor levels and therefore must lie in the vicinity of the acceptor levels in the *p*-type material. The voltage drop across the center of the junction is produced by a dipole layer composed of equal and opposite donor and acceptor charge, because free carriers are barred from this region by the difference of potential. This carrier-free region is known as the space-charge region, or depletion layer. We shall first determine the capacitance across this semi-insulating region, which is in series with the two low-resistance paths in the *n*- and *p*-type conducting regions. The potential barrier ϕ is of the order of the energy gap $\mathscr{E}_G$ which incidentally determines the spectral cutoff of the device as a photodiode. In the presence of an applied voltage, the barrier height becomes $(-V_A + \phi)$, where we define V_A as the applied bias in the *forward* direction. This is the direction that flattens the bands and causes easy flow, because the electrons may flow into the *p*-type material and vice versa. We may derive an expression for the width of the space-charge region by examination of Fig. 6.4. First, by charge neutrality, the area under the two charge distribution curves must be equal, so that

$$eN_A a = eN_D b \, .$$

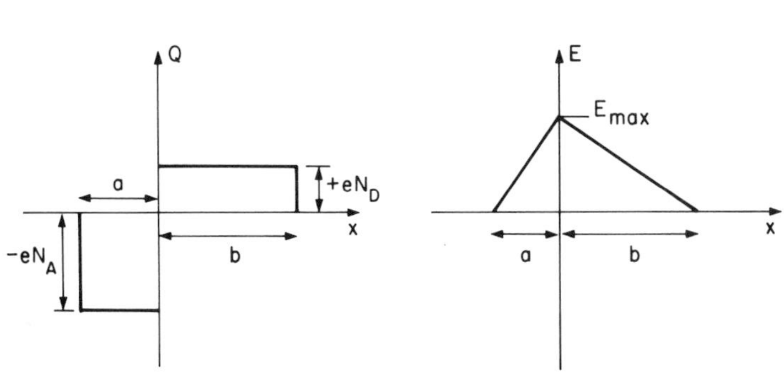

Fig. 6.4. Charge and field in space-charge region

Then, calculating the electric field, which must be zero at the edges of the space-charge region, we obtain

$$E_{\max} = \frac{eN_A a}{\kappa\varepsilon_0} = \frac{eN_D b}{\kappa\varepsilon_0}$$

from the relation between the included charge and the displacement, $\sigma = D = \varepsilon E$. Then the voltage may be obtained by integrating the field over the extent of the space-charge region, which is simply the area under the field curve. This is

$$(-V_A + \phi) = E_{\max}\left(\frac{a}{2} + \frac{b}{2}\right) = \frac{eN_A a}{2\kappa\varepsilon_0}(a + b)\,.$$

Now, obtaining the relationship between a and b from the charge-neutrality requirement, we find the width of the space-charge region,

$$w = \frac{\sqrt{2\kappa\varepsilon_0(-V_A + \phi)\left(1 + \dfrac{N_A}{N_D}\right)}}{eN_A}\,,$$

and the capacitance per unit area

$$\frac{C}{A} = \frac{\kappa\varepsilon_0}{w} = \sqrt{\frac{e\kappa\varepsilon_0}{2(-V_A + \phi)\,(N_A{}^{-1} + N_D{}^{-1})}}\,.$$

The latter expression may be proved by showing that the space-charge region behaves like a capacitor with spacing equal to the instantaneous value of the region's width. We shall used these expressions for width and capacitance later, in our treatment of the frequency response of the device. It should be noted that the capacitance decreases with increasing reverse bias, and that the heavier the doping, the larger the capacitance.

We now calculate the voltage-current characteristic for a *p-n* junction and we use the special configuration shown in Fig. 6.5, where one side of the junction is a thin layer of material configured to allow the radiation to produce hole-electron pairs near the space-charge region. We shall later describe how thin this layer should be to assure high quantum efficiency. If bias is applied to the junction, the minority-carrier density at the edge of the space-charge region may be written as

$$n_p = n_{p0}\, e^{eV_A/kT}\,; \quad p_n = p_{n0}\, e^{eV_A/kT}\,,$$

where n_{p0} and p_{n0} are equilibrium values.

The bias changes the barrier height; the resulting change of carrier density, because equilibrium is maintained with the majority carriers on the oppoiste side, is determined by the Maxwell-Boltzmann law (the low-occupation limit of

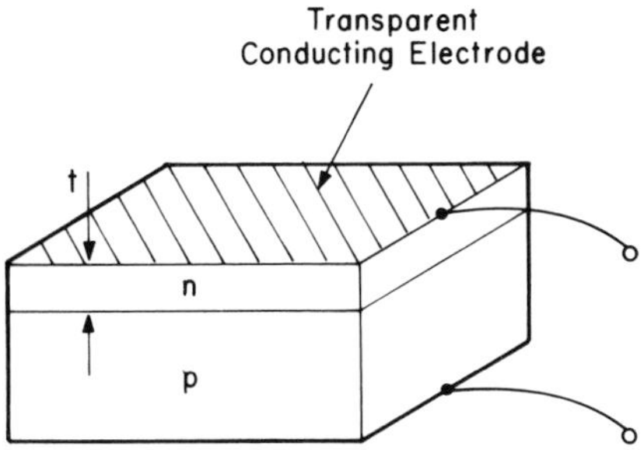

Fig. 6.5. Typical photodiode configuration

Fermi-Dirac statistics). This increased minority-carrier density causes diffusion away from the space-charge region; recombination occurs along the way. Fig. 6.6 shows schematically the carrier flow at either side of the junction. It should be noted that the surface of the thin n-type layer is treated as a reflector, and that the current flow, which is the negative gradient of the carrier density, becomes zero. The flow of the carriers is governed by the diffusion equations,

$$J_p = -eD_p \frac{dp}{dx}\,; \quad J_n = eD_n \frac{dn}{dx}\,,$$

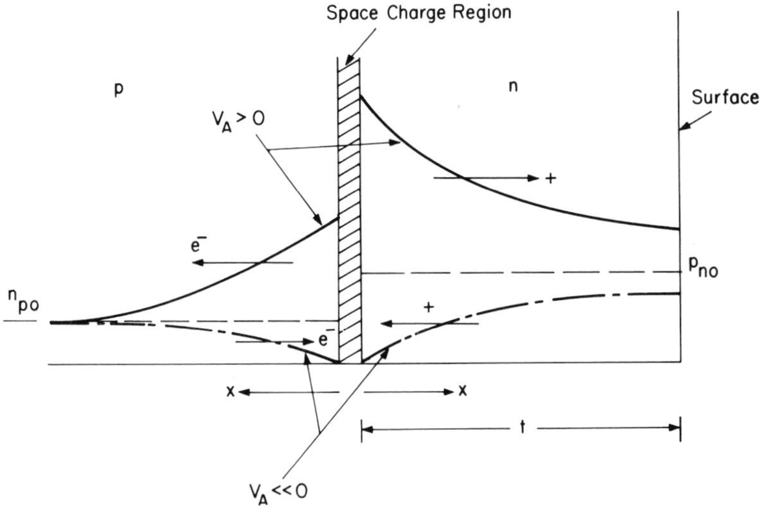

Fig. 6.6. Carrier flow in p-n junction

that is, the current density is proportional to the negative gradient of the carrier density, and

$$D_p \frac{d^2p}{dx^2} = -\frac{dp}{dt} = \frac{p_n - p_{p0}}{\tau} \; ; \quad D_n \frac{d^2n}{dx^2} = \frac{n_p - n_{p0}}{\tau} \, ,$$

which states that the rate of change of carrier flow is determined by the rate of recombination. The latter expression is simply a continuity equation for the carrier flow. The diffusion constant D is related to the mobility by the Einstein relation $D = (kT/e)\,\mu$, and has dimension $\mathrm{cm^2\,s^{-1}}$. The general solutions of the diffusion equation for holes and electrons are

$$p = p_{n0} + A\mathrm{e}^{+x/L_p} + B\mathrm{e}^{-x/L_p} \; ; \quad L_p = \sqrt{D_p \tau_p}$$

$$n = n_{p0} + C\mathrm{e}^{x/L_n} + D\mathrm{e}^{-x/L_n} \; ; \quad L_n = \sqrt{D_n \tau_n}$$

with the coefficients determined by appropriate boundary conditions. In the case of Fig. 6.6, the appropriate solutions are

$$n_p(x) = n_{p0} + n_{p0}\,(\mathrm{e}^{eV_A/kT} - 1)\,\mathrm{e}^{-x/L_n}$$

$$p_n(x) = p_{n0} + p_{n0}\,(\mathrm{e}^{eV_A/kT} - 1) \left[\frac{\mathrm{e}^{-x/L_p} + \mathrm{e}^{+(x-2t)/L_p)}}{1 + \mathrm{e}^{-2t/L_p}} \right] .$$

The hole-density expression indicates a symmetry that causes the gradient to become zero at $x = 1$. x is positive away from the junction in either direction; the minority carrier density at $x = 0$ agrees with the value previously given for the edge of the space-charge region.

The total current flow in the device is equal to the flow across the space-charge region, which is given by the individual diffusion currents at the edge of the region. Using the expression for the current density evaluated at $x = 0$, we obtain finally

$$J = \left(\frac{eD_n n_{p0}}{L_n} + \frac{ep_{n0}t}{\tau_p} \right) (\mathrm{e}^{eV_A/kT} - 1) \, ,$$

where, in the case of the hole current, we have assumed that $t \ll L_p$. The latter term in the hole current, subject to this restriction, is simply the total generation of carriers in the thin layer of thickness t. The total current for large reverse bias V_A is

$$i = JA = A \left(\frac{eD_n n_{p0}}{L_n} + \frac{ep_{n0}t}{\tau_p} \right) \triangleq i_D \, ,$$

which is known as the saturation current. We shall use the expression, dark current, because it is the component of current flow that is present in the absence of any illumination. We now hypothesize that radiation strikes the device in the vicinity of the junction and produces extra hole-electron pairs. The minority carriers from such a process may diffuse to the junction and cause a current whose direction is the same as the reverse or saturation current. Later, we shall calculate the effective quantum efficiency for this process. We may write the current due to signal and background power as

$$i = \frac{\eta e(P_S + P_B)}{h\nu},$$

where the value of η is determined by the proximity to the space-charge region of the pair-creation process.

We now consider the noise associated with the photodiode, which is similar to that in a vacuum photodiode except that we must account for current flowing in *both* directions across the space-charge region. Specifically, under bias or even at zero bias, there are always several current components flowing across the junction. These include the dark current i_D, the photoinduced current, and a forward current $i_D \exp(eV_A/kT)$. Each of these currents produces shot noise, and the noises are statistically independent. Therefore the mean-square noise current

$$\overline{i_N^2} = 2e\left[i_D\, e^{eV_A/kT} + i_D + \frac{\eta e(P_S + P_B)}{h\nu}\right].$$

It is interesting to consider the noise current at zero bias, in the absence of radiation. This is

$$\overline{i_N^2} = 4ei_D B,$$

and if we note that the resistance of the junction at zero bias is given by

$$\frac{1}{R} = \left(\frac{di}{dV}\right)_{V_A = 0} = \frac{ei_D}{kT},$$

we obtain for the mean-square noise current

$$\overline{i_N^2} = 4e\left(\frac{kT}{eR}\right)B = \frac{4kTB}{R},$$

which is just Johnson noise, as would be expected for a system in thermal equilibrium.

Using (4.1) for the signal-to-noise ratio, we may now write the appropriate expression for the reverse-biased photodiode, using

$$\overline{i_S^2} = \eta^2 e^2 P_S^2/(h\nu)^2$$

$$i_N^2 = 2e\left[i_D + \frac{\eta e(P_S + P_B)}{h\nu}\right]B,$$

with the result

$$\frac{S}{N} = \frac{\eta P_{\mathrm{S}}^2}{2h\nu B\left(P_{\mathrm{S}} + P_{\mathrm{B}} + \dfrac{h\nu i_{\mathrm{D}}}{\eta e}\right)\left[1 + 2\left(\dfrac{h\nu}{e}\right)\left(\dfrac{kT_{\mathrm{N}}}{e}\right)\left(\dfrac{1}{\eta R}\right)\left(\dfrac{1}{P_{\mathrm{S}} + P_{\mathrm{B}} + \dfrac{h\nu i_{\mathrm{D}}}{\eta e}}\right)\right]}, \tag{6.8}$$

which is identical with the result obtained for the vacuum photodiode, (5.1).

Before treating the quantum efficiency, we complete our diode description by discussing the photovoltaic mode, in which the diode is operated without bias into a high-impedance load. This mode suffers in sensitivity, as we shall see, but is used in some cases in which the reverse-leakage resistance of the junction is so low that applied bias causes excess noise currents that preclude optimum performance. The production of hole-electron pairs near the junction results in diffusion of minority carriers to the space-charge region, which produce excess current in the *reverse* direction. Thus, if the diode is open circuited, this carrier flow will be counteracted by a terminal voltage in the *forward* direction, which maintains net zero current flow. We may therefore write

$$i_{\mathrm{S}} + i_{\mathrm{B}} = \frac{\eta e(P_{\mathrm{S}} + P_{\mathrm{B}})}{h\nu} = i_{\mathrm{D}}(\mathrm{e}^{eV/kT} - 1),$$

and the induced voltage becomes

$$V = \frac{kT}{e} \ln\left(\frac{i_{\mathrm{B}} + i_{\mathrm{S}} + i_{\mathrm{D}}}{i_{\mathrm{D}}}\right).$$

Now for small-signal operation, the signal voltage is given by

$$v_{\mathrm{S}} = \frac{dV}{di_{\mathrm{S}}}\, i_{\mathrm{S}} = \frac{kT}{e}\left(\frac{i_{\mathrm{S}}}{i_{\mathrm{S}} + i_{\mathrm{B}} + i_{\mathrm{D}}}\right),$$

where the equation applies for $v_{\mathrm{S}} \ll kT/e$. The noise current is the shotnoise associated with the total current flow in both directions, and the current in *either* direction is $(i_{\mathrm{S}} + i_{\mathrm{B}} + i_{\mathrm{D}})$. The mean-square noise current is therefore

$$\overline{i_{\mathrm{N}}^2} = 4e(i_{\mathrm{S}} + i_{\mathrm{B}} + i_{\mathrm{D}})\, B.$$

A further important property of the device is its resistance at the operating point, which is given by

$$\frac{1}{R_{\mathrm{D}}} = \frac{d}{dV}[i_{\mathrm{d}}(\mathrm{e}^{eV/kT} - 1)] = \frac{ei_{\mathrm{D}}}{kT}\mathrm{e}^{eV/kT} = \frac{e}{kT}(i_{\mathrm{S}} + i_{\mathrm{B}} + i_{\mathrm{D}}). \tag{6.9}$$

Although we have treated the device as an open-circuit voltage source, we may use the current-source equivalent if we realize that the current is shunted by the diode resistance and any external load R_L, due to leakage or required for higher frequency response. The resultant currents are

$$\overline{i_S^2} = \eta^2 e^2 P_S^2/(h\nu)^2$$

$$\overline{i_N^2} = 4e\left[\frac{\eta e(P_S + P_B)}{h\nu} + i_D\right]B\,,$$

and the final signal-to-noise expression is

$$\frac{S}{N} = \frac{\eta P_S^2}{4h\nu B\left(P_S + P_B + \dfrac{h\nu i_D}{\eta e}\right)\left[1 + \left(\dfrac{h\nu}{e}\right)\left(\dfrac{kT_N}{e}\right)\left(\dfrac{1}{\eta R_L}\right)\left(\dfrac{1}{P_S + P_B + \dfrac{h\nu i_D}{\eta e}}\right)\right]}\,. \tag{6.10}$$

Note that in (6.10), R_D is not included as a contributor to the Johnson noise, because its contribution is represented in the shot-noise-current term, as demonstrated in our zero-bias calculation of junction noise. The equation may be written in a more understandable form if we note that the last term in the denominator contains R_D, as derived above. Taking this into account, we may write the simpler expression,

$$\frac{S}{N} = \frac{\eta P_S^2}{4h\nu B\left(P_S + P_B + \dfrac{h\nu i_D}{\eta e}\right)\left[1 + \left(\dfrac{R_D}{R_L}\right)\left(\dfrac{T_N}{T_D}\right)\right]}\,,$$

where T_D is the physical temperature of the diode. Our result does not differ drastically from the reverse-biased case except for the factor-of-two reduction of ultimate signal-to-noise ratio because of the doubled shot-noise contribution. The reduced sensitivity is somewhat akin to that of the photoconductor where in the latter, the random recombination adds excess noise although, in the former, the random replacement of carriers across the junction provides the excess contribution.

We now treat the behavior of the quantum efficiency by examining how close to the space-charge region the radiation must produce hole-electron pairs for efficient collection. We start by writing the general diffusion equation, including time-varying carrier densities, because we wish to determine the effect of the optical modulation frequency. This equation for holes is

$$\frac{dp}{dt} = D_p \frac{d^2p}{dx^2} - \frac{p}{\tau}$$

(where p is the excess hole density), which simply expresses the conservation of carrier density in a small increment of length. That is, the rate of increase of carrier density is equal to the inward flow minus the outward flow minus the recombination, because the second-derivative term is simply the divergence of the diffusion current. The general solution to this equation is similar to that derived above for the non-time-dependent case; specifically, if we assume an excess carrier density of the form

$$p = \mathrm{Re}\{p_0\, \mathrm{e}^{\pm x/L}\, \mathrm{e}^{\mathrm{i}\omega t}\},$$

then the solution is

$$p = \mathrm{Re}\{A\mathrm{e}^{+x/L} + B\mathrm{e}^{-x/L}\},$$

where the quantity L, which we call the complex diffusion length, is given by

$$L = \sqrt{\frac{D_\mathrm{p}}{\mathrm{i}\omega + \dfrac{1}{\tau_\mathrm{p}}}}\,.$$

Thus, the carrier density in the presence of an ac excitation decays with attenuation and phase shift that are determined by the complex diffusion length L. We now treat the collection efficiency of a junction with a sheet source of radiation spaced a distance d from the space-charge region. As depicted in Fig. 6.7, the requirement for zero excess minority-carrier density at the space-charge region is satisfied by a negative-hole distribution mirror imaged about $x = 0$. We may write the carrier distribution as

$$p = p_0\,[\mathrm{e}^{-(x+d)/L} - \mathrm{e}^{+(x-d)/L}]$$

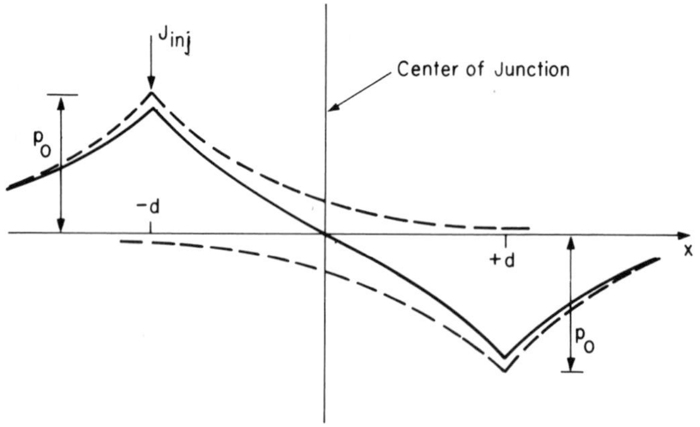

Fig. 6.7. Calculation of collection efficiency

for the region between $-d$ and $+d$. The carrier current that flows to the right in this region then becomes

$$J_{\mathrm{R}}(x > -d) = \frac{eD_{\mathrm{p}}p_0}{L}\,[\mathrm{e}^{-(x+d)/L} + \mathrm{e}^{+(x-d)/L}]\,,$$

which at $x = -d$, the current-injection point, is

$$J_{\mathrm{R}}(x = -d) = \frac{eD_{\mathrm{p}}p_0}{L}\,(1 + \mathrm{e}^{-2d/L})\,.$$

Similarly, there is a current flowing to the left at $x = -d$ with the same sign for the first term and opposite sign for the second, because the negative mirror-image current is continuous. This current is

$$J_{\mathrm{L}}(x = -d) = \frac{eD_{\mathrm{p}}p_0}{L}\,(1 - \mathrm{e}^{-2dL})\,;$$

therefore tht total injected-current density is

$$J_{\mathrm{inj}} = J_{\mathrm{R}}(x = -d) + J_{\mathrm{L}}(x = -d) = \frac{2eD_{\mathrm{p}}p_0}{L}\,.$$

The current collected at the junction is given by J_{R} evaluated at $x = 0$, and the collection efficiency is

$$\frac{J_{\mathrm{coll}}}{J_{\mathrm{inj}}} = \mathrm{e}^{-d/L}\,.$$

When the argument of the exponential is rewritten for the low-frequency and high-frequency cases, the answer is

$$\frac{J_{\mathrm{coll}}}{J_{\mathrm{inj}}} = \mathrm{e}^{-d/\sqrt{D\tau}}\,; \quad \omega \ll \frac{1}{\tau} := \mathrm{e}^{-(1+\mathrm{i})\frac{d}{\sqrt{2D/\omega}}}\,,\ \omega \gg \frac{1}{\tau}\,.$$

The actual injection or generation rate is dependent upon the absorption process as the radiation penetrates the device normal to the junction, as in the structure of Fig. 5.5. There are two modes of operation of such a device. In one, most of the absorption takes place in the thin n layer on the surface. In the other, the surface-layer absorption is negligible and the carriers are mainly produced in the underlying p-type material. The latter case may be due to a very small value of t, the layer thickness, or the energy gap of the top layer may be wider and thus closer to the cutoff frequency than the bulk material, because of heavier (degenerate) doping or a different alloy composition. The case for absorption in the surface layer is difficult to calculate exactly, because some absorption always occurs in the bulk material, at a different rate and with a different value of L, and also the surface introduces an additional boundary condition that is not considered in the foregoing derivation. Suffice it to say, the quantum efficiency approaches $(1 - R)$, where R is the optical reflectance, if the thickness t is much less than $1/\alpha$, where α is the absorption coefficient, and t is much less than L, the complex diffusion length.

For the case of negligible absorption in the surface layer, the collection efficiency may be found in a straightforward manner, by integrating the efficiency over the distribution of generated current. The appropriate expression is

$$\eta = (1 - R) \int_0^\infty e^{-\alpha x} \cdot e^{-x/L}\, dx = (1 - R)\left(\frac{\alpha}{\alpha + 1/L}\right) ;$$

thus, the quantum efficiency approaches $(1 - R)$ when α is much greater than the inverse of the diffusion length.

Because photodiodes use the intrinsic mode of photoexcitation, their wavelength of operation is determined by the energy gap of the semiconductor. Typical materials and wavelengths are germanium, 1.8 μm; silicon, 1.3 μm; and indium antimonide, 5 μm, as well as a newer family of mixed alloys whose gap may be varied by varying the ratio of two of the major components. These include $Pb_{(1-x)}Sn_xTe$, $Pb_{(1-x)}Sn_xSe$, and $Hg_{(1-x)}Cd_xTe$. As an example of device behavior, we shall calculate the performance of a mercury cadmium telluride diode designed to operate at 10 μm (*Spears,* 1977). The parameters of the device are

$$\begin{aligned} \mathscr{E}_G &= 0.1\text{V} \approx \phi & \tau_n &= \tau_p = 10^{-7}\ \text{s} \\ \kappa &= 18 & n_i &= 10^{13}\ \text{cm}^{-3} \\ N_D &= 10^{15}\ \text{cm}^{-3} & \alpha &= 4 \times 10^3\ \text{cm}^{-1} \ @\ 10\ \mu\text{m} \\ N_A &= 10^{17}\ \text{cm}^{-3} & n_{p0} &= \frac{n_i^2}{N_A}\ ;\ p_{n0} = \frac{n_i^2}{N_D} \\ \mu_n &= 10^5\ \text{cm}^2\ \text{V}^{-1}\ \text{s}^{-1} & & \\ \mu_p &= 10^2\ \text{cm}^2\ \text{V}^{-1}\ \text{s}^{-1}, & & \end{aligned}$$

with all values applicable at $T = 77$ K. We assume a thin n-type layer 5 μm thick on top of the bulk p-type material and a square 0.2 mm on a side. In the n-type material, the absorption coefficient will be much less than the given value because of conduction band filling caused by the extremely small effective mass of an electron, as a result of which the majority of photoinduced carriers are produced in the p-type material. This, of course, is advantageous, because high electron mobility contributes to high efficiency of collection. The applied voltage in these devices is limited to 2 V because of extra leakage current associated with the edge of the junction region. In fact, typical reverse resistances are the order of 500 Ω. Calculating the width of the space-charge region, we obtain

$$w = \sqrt{\frac{2\kappa\varepsilon_0(-V_A + \phi)}{eN_D}} = 2 \times 10^{-4}\ \text{cm} = 2\ \mu\text{m}$$

because $N_D \ll N_A$. The capacitance then becomes

$$C = \frac{\kappa\varepsilon_0 A}{w} = 3 \times 10^{-12}\ \text{F} = 3\ \text{pF}\,.$$

Using D_n of 600 cm^2 s^{-1}, from the expression $D = kT\mu/e$, we obtain for the diffusion length,

$$L_n = \sqrt{D_n\tau_n} = 8 \times 10^{-3}\ \text{cm} = 80\ \mu\text{m}\ ;$$

at low frequencies, the quantum efficiency is roughly $(1 - R)$, because the absorption length $1/\alpha$ is only 2.5 μm. The high-frequency cutoff occurs at $1/\alpha$ approximately equal to $(D/\omega_c)^{1/2}$, so that

$$f_c = \frac{\alpha^2 D}{2\pi} = 1600\ \text{MHz}\,.$$

In practice, the cutoff frequency will be limited by the junction capacitance; for a load determined by the leakage resistance of 500 Ω the cutoff occurs at

$$f_c = \frac{1}{2\pi RC} = 100\ \text{MHz}\,.$$

Calculating the dark current, we obtain

$$i_D = \left(\frac{ep_{n0}t}{\tau_p} + \frac{eD_n n_{p0}}{L_n}\right) A = 20 \times 10^{-9}\ \text{A}$$

and ask whether the *NEP* is dark-current or amplifier-noise limited. The appropriate expressions for the two cases are

$$(NEP)_{DL} = \frac{h\nu}{e}\sqrt{2ei_d B} = 1.7 \times 10^{-14}\sqrt{B}\ \text{W}$$

$$(NEP)_{AL} = \frac{2h\nu}{e}\sqrt{\frac{kT_N B}{R}} = 5 \times 10^{-13}\sqrt{B}\ \text{W with } T_N = 300\ \text{K}\,,$$

so we see that the device is strongly amplifier-noise limited. The value of D^* is given by

$$D^* = \frac{\sqrt{AB}}{NEP} = \frac{2 \times 10^{-2}}{5 \times 10^{-13}} = 4 \times 10^{10}\ \text{cm} \cdot \text{Hz}^{1/2}\ \text{W}^{-1}$$

which is not much greater than the limiting value for a full 180° field of view at $T = 300$ K for a 10 μm detector. It should be noted that the leakage resistance is the limiting factor in the sensitivity, because the device would approach dark-current-limited operation if the load resistance could be increased by a factor of about 10^4 or to 5 MΩ. The frequency response would of course be limited by the capacitance to 10 kHz, which is still quite adequate for many incoherent detection systems.

The diode we have described is especially useful for heterodyne detection, for which the local-oscillator current can be made high enough so that the detector becomes shot-noise limited. In this case, the full frequency response may be used by operating with a 50 Ω load resistance. In this application, we should also note the cutoff frequency associated with the transit time across the space-charge region. This time is *not* determined by the mobility because the fields in the space-charge region are high enough so that the electron velocity is limited to the order of 10^7 cm s^{-1} by interaction with lattice phonons. This saturation velocity yields a transit time for the 2 μm space-charge width of 2×10^{-11} s, corresponding to a cutoff frequency well above the diffusion-limited value.

6.3 Avalanche Photodiodes

The analog of the photomultiplier in the case of a semiconductor is the avalanche photodiode, in which the photoexcited carrier produces extra hole-electron pairs as it moves through the space-charge region. The pair creation occurs at electric fields of the order of 10^5 V cm^{-1} and is characterized by the ionization coefficient, α for electrons and β for holes. The coefficients α and β vary rapidly with electric field according to a relation of the form $\exp(-E_0/E)$. We shall not be concerned with the exact field dependence of the coefficients, but shall treat two simple cases, that where $\beta = 0$ and the other case where $\alpha = \beta$. From the definition of the ionization coefficients, we may write the two equations for the electron and hole current,

$$\frac{dI_n}{dx} = \alpha I_n + \beta I_p \,, \tag{6.11}$$

$$\frac{dI_p}{dx} = -\alpha I_n - \beta I_p \,, \tag{6.12}$$

where x is the distance through the space-charge region as defined in Fig. 6.8. Considering first the case of $\beta = 0$, we assume the hole current entering from the right to be zero, and the electron current leaving the left side (that is, electrons moving to the right) as $I_n(0)$. By charge neutrality, the current must be continuous. Therefore, we may write

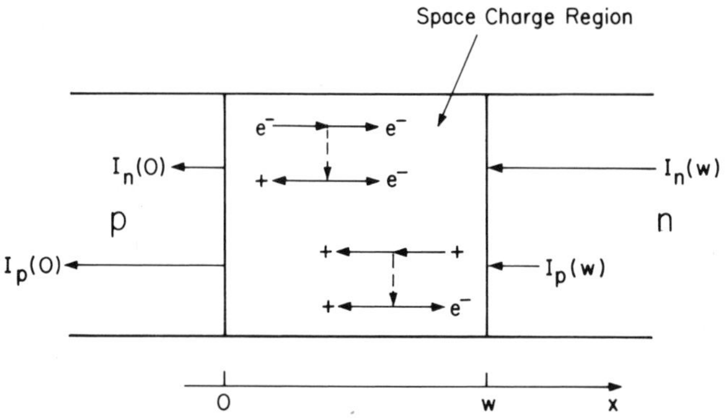

Fig. 6.8. Hole and electron currents in avalanche photodiode

$$I = I_n(w) = I_n(x) + I_p(x)\,.$$

Solving (6.11) for I_n, with $\beta = 0$, we obtain in successive steps

$$\frac{dI_n}{dx} = \alpha I_n\,,$$

$$\int_{I_n(0)}^{I_n(x)} \frac{dI_n}{I_n} = \int_0^x \alpha dx\,,$$

$$I_n(x) = I_n(0)\,\mathrm{e}^{\int_0^x \alpha dx}\,,$$

where we note that α is a function of x, as determined by the electric field. Because of the rapid variation of α with field, it is simpler conceptually to introduce the variable,

$$y = \int_0^x \alpha dx\,,$$

which results in the plot of Fig. 6.9. Here note that the hole current $I_p(x)$ is given by

$$I_p(x) = I_n(w) - I_n(x) = I - I_n(x)$$

from the continuity relationship. The variation of total current I with voltage is quite complicated since the function y increases rapidly with increasing voltage; however, in the case of $\beta = 0$, the current always remains finite. In the case of a finite value for β, we shall see that the current can become infinite

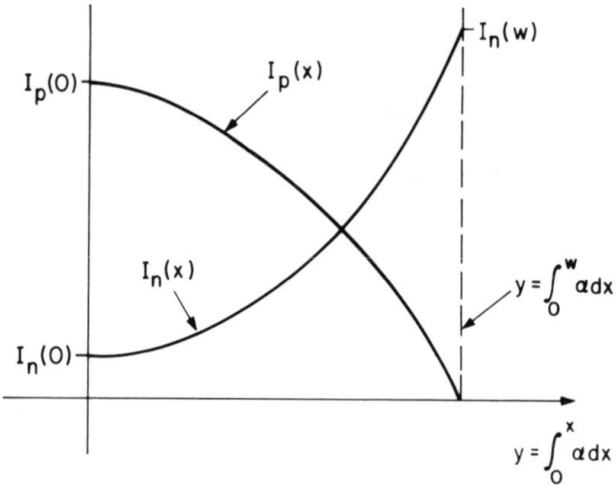

Fig. 6.9. Current variation for $\beta = 0$

through a process called avalanche breakdown. To understand this case, we shall treat the simple problem of $\alpha = \beta$. To do so, using the master equations combined with the continuity requirement, we obtain:

$$\frac{dI_n}{dx} = \alpha I_n + \beta I_p = \alpha I_n + \beta(I - I_n)$$
$$= (\alpha - \beta)\, I_n + \beta I = (0) I_n + \beta I = \alpha I\,.$$

Because I is independent of x, we write

$$\int_{I_n(0)}^{I_n(x)} dI_n = I \int_0^x \alpha dx: \quad I_n(x) - I_n(0) = I \int_0^x \alpha dx\,,$$

and solving for I in terms of $I_n(0)$, we obtain

$$I = I_n(w) = \frac{I_n(0)}{1 - \int_0^w \alpha dx}\,, \qquad (6.13)$$

which becomes infinite if the integral becomes unity. This represents avalanche breakdown. For the case of unequal ionization coefficients, the behavior is more complicated but still yields a breakdown condition. The behavior of the current under these conditions is given in Fig. 6.10 again plotted against the variable y.

We shall now treat the noise process for these two limiting cases, because they represent the extremes of noise performance for an avalanche photodiode, as we shall see.

The noise process in an avalanche photodiode is similar in many ways to that in the photomultiplier. We define a current gain M for the full device, which

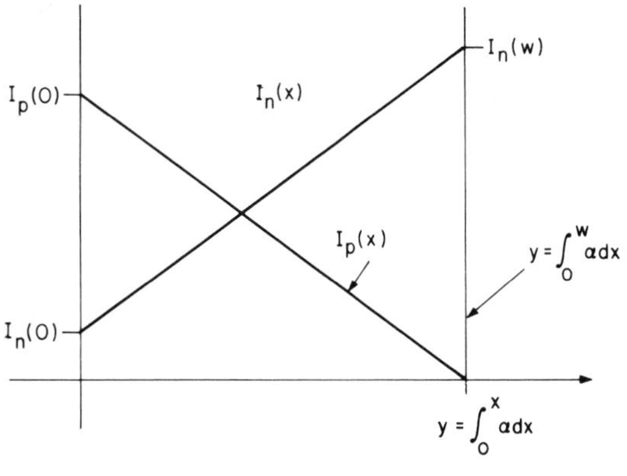

Fig. 6.10. Current variation for $\alpha = \beta$

is simply $I/I_n(0)$, and we also define the current gain $M(x)$, which is the change of the output current I due to small fluctuation of the current $I_n(x)$. We thus write the mean-square noise current as

$$\overline{i_N^2} = 2eB\left[M^2\, I_n(0) + \int_{I_n(0)}^{I} M^2(x)\, dI_n(x)\right], \tag{6.14}$$

which states that the increase of current dI_n at position x produces pure shot noise of form $2eBdI_n$, and that this noise is in turn multiplied by the square of the remaining current gain from position x to the output at $x = w$. The assumption of pure shot noise means that the ionization process is completely random, or that the probability of ionization in a finite distance is Poisson distributed. The multiplication factor $M(x)$ is simply the ratio $I/I_n(x)$, because from our original derivation we may write

$$I_n(w) = I_n(x)\, e^{\int_x^w \alpha dx},$$

and the exponential term is the current gain from x to $x = w$. Substituting for this gain term in the noise equation leads to,

$$\begin{aligned}\overline{i_N^2} &= 2eB\left[M^2\, I_n(0) + \int_{I_n(0)}^{I_n(w)} \frac{I_n^2(w)}{I_n^2(0)}\, dI_n(x)\right]\\ &= 2eB\left\{M^2 I_n(0) + I^2\left[\frac{1}{I_n(0)} - \frac{1}{I_n(w)}\right]\right\}\\ &= 2eB\, I_n(0)\,(M^2 + M^2 - M) = 2eM^2\, I_n(0)\, B\left(2 - \frac{1}{M}\right)\end{aligned}$$

and we note that for high gain, the output noise power is twice that for an ideal multiplier. This is not too surprising if we refer back to the photomultiplier case for Poisson-distributed secondary emission, (5.3). If we think of the avalanche process as equivalent to a set of dynodes with $m = 2$, we again obtain the same noise factor of two.

We now treat the noise for the case of equal ionization coefficients and shall find a much noisier behavior, which is associated with the much more complicated amplification process in which two types of carrier produce current amplification. We shall use the same expression (6.14) for the noise in terms of the multiplication factor as a function of distance through the space-charge layer, but the behavior of the quantity $M(x)$ is markedly different in this case. To derive the multiplication term, we ask for the change of output current $I = I_n(w)$ that is associated with an incremental increase of current $\Delta I_n(x)$ at x. Using the equations derived for $\alpha = \beta$, we write

$$I_n(x) - I_n(0) = I_n(w) \int_0^x \alpha dx$$

$$I_n(w) - [I_n(x) + \Delta I_n(x)] = I_n(w) \int_x^w \alpha dx .$$

Adding the two equations yields

$$I_n(w) - I_n(0) - \Delta I_n(x) = I_n(w) \int_0^w \alpha dx,$$

from which the output current is given by

$$I_n(w) = \frac{I_n(0) + \Delta I_n(x)}{1 - \int_0^w \alpha dx} .$$

This is simply the regular amplification equation. A small increase of current *anywhere* in the space-charge region increases the output current by the same factor, namely, the gain M of the whole structure. This can be understood physically by realizing that the small current increase sends extra electrons to the right but an equal amount of holes to the left; the sum of these two groups of particles produces a net increase of current, independent of the location of the initial fluctuation.

Using this result for the multiplication factor yields for the noise current,

$$\begin{aligned} \overline{i_N^2} &= 2eB\left[M^2 I_n(0) + \int_{I_n(0)}^{I_n(w)} M^2 dI_n(x)\right] \\ &= 2eB\{M^2 I_n(0) + M^2[I_n(w) - I_n(0)]\} = 2eM^3 I_n(0)B . \end{aligned} \tag{6.15}$$

We see that the effective noise factor for the case of equal ionization coefficients is equal to M, rather than 2, as in the previous case. Thus the signal-to-noise ratio of an avalanche diode increases to an optimum value as the ionization coefficient for the *injected* or photoinduced carrier becomes large compared to that of the opposite carrier type. In actual practice, carefully designed diodes have been shown to exhibit a mean-square noise current that behaves as $M^{2.1}$, rather than M^2 as the theory would indicate. Details of typical structures and their measured behaviors are discussed by *Melchior* et al. (1970). Unfortunately, minute nonuniformities in the junction structure prevent operation of avalanche diodes at gains much above $M = 100$, because a small region with slightly higher gain will cause avalanche breakdown or excess constricted current flow, which results in rapid deterioration of performance.

In conclusion, we should point out that the noise caused by injection of the *wrong* carrier is even worse than described. In fact, as discussed in detail by *McIntyre* (1966), who derived the original noise theory, the noise factor approaches $M\ \beta/\alpha$ for the structure we have considered.

The frequency response of the diode is also a strong function of the ratio α/β. *Emmons* (1967) has shown that for $\alpha \gg \beta$, the frequency response is just the inverse of the transit time, just as in a standard photodiode. For $\alpha = \beta$, however, the response is reduced by a factor M, because the effective storage time becomes equivalent to M transit times across the space-charge region. Present avalanche photodiodes are limited to silicon (1.3 μm) and gallium arsenide (0.9 μm), although some progress has been reported for indium antimonide (5 μm).

Problems

6.1 In Chapter 2, we derived the background-limited detectivity for a photodetector exposed to a 300 K field of view of full cone angle 180°. The sensitivity was found to increase exponentially at the shorter wavelengths but the detector, of course, would have a large increase of resistance, and amplifier noise would eventually limit the ultimate sensitivity. Now consider a photoconductor 1 mm square exposed to a 60° full cone angle 300 K field of view. If the detector load resistance is limited to 1 MΩ, at what *wavelength* does the detector become amplifier noise limited, assuming $\eta = 1$, and $G = 1$, and $T_N = 300$ K. Repeat for a 100 MΩ resistance.

6.2 What is the *NEP* at this first wavelength, taking into account both the amplifier and background noise.

6.3 Find the value of D^* *neglecting amplifier noise.*
(Assume that the detector and cavity temperature are low enough so that the dark conductance and local radiation are negligible).

6.4 A silicon avalanche diode with $\alpha/\beta \gg 1$ has a quantum efficiency η of 50%

and is operated at 1.06 μm with a current multiplication factor $M = 10^3$, a 50 Ω load, and a 3 dB noise figure amplifier. For bandwidths less than 1000 MHz, is the device amplifier- or signal-noise limited? What is the *NEP* in terms of bandwidth *B*.

6.5 Obtain a similar expression, i.e., *NEP* as a function of bandwidth for a photomultiplier with extremely high gain, a 2% quantum efficiency, and a noise factor $\Gamma = 2$, neglecting dark current and background.

6.6 Find the *NEP* for each of the above for 20 ns pulses. Assume $B =$ the inverse of the pulsewidth.

7. Thermal Detection

We now consider the detection of optical and infrared radiation by means of the temperature change of the detecting element, rather than by use of the direct photoexcitation process. Because detectors of this type have no inherent long-wavelength cutoff and the sensitivity limitation is mainly due to the thermal background power, we shall first investigate in detail the fluctuations of the thermal radiation field. To this point, we have assumed that the thermal background may be treated as a constant power striking the detector, but we have noted that this assumption is valid only for Bose-Einstein state occupancy much less than unity. In the first part of the chapter, we shall derive the effective fluctuation of the radiation field in the general case in which f, the occupancy factor, may have any value. Our treatment also yields the basis for thermal or Johnson noise.

7.1 Fluctuations of the Radiation Field

We start by considering a detector exposed to thermal radiation that originates at a large distance from the detector so that all wavefronts from the radiating medium are parallel and thus that all the field components are coherent over the detector surface. This is equivalent to stating that the source is in the far field of the detector and that all radiators within the source lie within one spatial mode of the detector. We now wish to determine the probability distribution of the instantaneous power that strikes the detector surface, assuming random spontaneous emission from the ensemble of radiators in the emitting medium. Because each atom or molecule in the ensemble emits with random phase and random polarization, the instantaneous electric field at any point on the detector surface may be described by the expression

$$E = E_1(t) \sin 2\pi\nu_0 t + E_2(t) \cos 2\pi\nu_0 t;$$

this expression holds for each of the polarizations for the incoming wave, and ν_0 is the center optical frequency at which the observation is made. Because E is a random variable with zero mean, it may be shown to vary in a Gaussian manner. From a statistical point of view, E is distributed in a complex Gaussian manner, a property first deduced by Rayleigh. Schematically, as in Fig. 7.1, the two quadrature components of the field arise by the random summation of the small fields from each of the independent radiators. This behavior is akin to

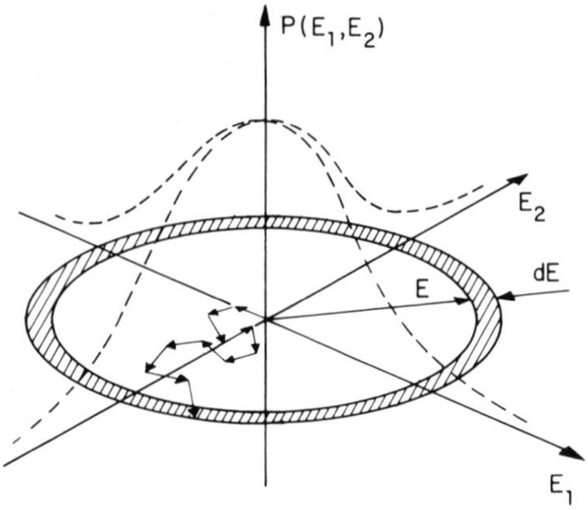

Fig. 7.1. Probability density for E in complex Gaussian noise

random walk in the two-dimensional plane. The probability distribution is Gaussian in both E_1 and E_2 and the probability density in E space is

$$p(E) \propto E \exp[-(\overline{E_1^2})/(E^2)] \exp[-(E_2^2)/(\overline{E^2}) = E \exp[-(E^2)/(\overline{E^2})],$$

because the probability of a field value between E and $E + dE$ is proportional to the area inside the annulus of radius E. The probability density for the instantaneous power may be written

$$p(P)\, dP = p(E)\, dE,$$

where dP is the corresponding increment of power associated with a change dE of the field. But dP is proportional to EdE, because the power is quadratic in E; therefore

$$p(P)\, EdE \propto E\, \mathrm{e}^{-(E^2)/(\overline{E^2})}\, dE$$

and

$$p(P) \propto \mathrm{e}^{-(E^2)/(\overline{E^2})} = \mathrm{e}^{-P/\bar{P}},$$

where $\bar{P}$ is the average power. The coefficient in the proportionality is simply $1/\bar{P}$, because

$$\bar{P} = \int_0^\infty p(P)\, PdP = \int_0^\infty \frac{P}{\bar{P}} \mathrm{e}^{-P/\bar{P}dP=\bar{P}} \int_0^\infty x\mathrm{e}^{-x}\, dx = \bar{P}.$$

Note that, at this stage, we have not examined the time variation of the power fluctuation. Thus far we have specified only that the two quadrature components of the field have amplitudes that are Gaussian distributed. We shall investigate the temporal behavior of the amplitudes later in the derivation. We now wish to determine the mean-square fluctuation of the power, which is defined by

$$\overline{\Delta P^2} = \overline{(P - \bar{P})^2} = (\overline{P^2 - 2P\bar{P} + \bar{P}^2}) = \overline{P^2} - \bar{P}^2 .$$

Solving for the mean-square power, we obtain

$$\overline{P^2} = \int_0^\infty p(P)P^2\, dP = \int_0^\infty \frac{P^2}{\bar{P}} e^{-P/\bar{P}}\, dP = \bar{P}^2 \int_0^\infty x^2 e^{-x} dx = 2\bar{P}^2 ,$$

and thus

$$\overline{\Delta P^2} = \overline{P^2} - \bar{P}^2 = \bar{P}^2 , \tag{7.1}$$

which states that the *mean-square fluctuation* is the *square* of the *mean*. This very important classical result is the basis for thermal or Johnson noise, because for very low frequencies, for which $h\nu \ll kT$, the thermal power per mode becomes $kT\Delta\nu$ or kTB (see Sec. 3.4 and Problem 1.4). The associated *noise* expressed as the *rms* power fluctuation is equal to the mean or kTB. Because a two-wire circuit or transmission line supports only one spatial mode (typically a TEM mode), (7.1) is the appropriate expression for a circuit element, transmission line or single-mode waveguide.

We now expose our hypothetical photon detector to a blackbody radiation field that comes from all directions, so that the radiation that strikes the detector is distributed through many separate spatial modes of space. We shall then set out to determine the mean-square noise current in the detector and from this deduce the *effective* power fluctuation that strikes the detector surface. The noise current from an ideal photodetector with quantum efficiency η is composed of two terms; first, the "shot noise" associated with the detection process and second, a current fluctuation that is associated with the mean-square power fluctuation. We thus write, using (2.10) and (2.11),

$$\overline{i_N^2} = 2e\frac{\eta e\bar{P}}{h\nu} B + \left(\frac{\eta e}{h\nu}\right)^2 \overline{\Delta P^2} .$$

We now must find the power fluctuation that is associated with a large number of modes. We refer to our previous derivation where we assumed that the electric fields were spatially coherent over the full detector surface. This assumption is valid only for the radiating source contained in the angular subtense of a single mode. In fact, the fluctuation of the total power that strikes the detector is given by the mean-square fluctuations of each mode, summed over all modes. This is true because the product of the electric fields in two separate modes integrated over the surface area is zero (see Chap. 3) and there is no contribution to either the power or the power fluctuation. Therefore,

$$\overline{\Delta P^2} = \sum_k (\overline{\Delta P^2})_k = \sum_k \bar{P}_k^2 ,$$

where $\bar{P}_k$ is the power per mode and k is the mode number. Thus, the expression for the noise current is

$$\overline{i_N^2} = 2e \frac{\eta e}{h\nu} B\bar{P} + \left(\frac{\eta e}{h\nu}\right)^2 \sum_k \bar{P}_k^2 , \tag{7.2}$$

where

$$\bar{P} = \sum_k \bar{P}_k .$$

What we are saying is that the field in each mode beats with itself, which causes the fluctuation of detector current, but that fields of separate modes do not interact.

To calculate the spectral density of the noise current, we assume that the optical bandwidth of the radiation that strikes the detector is limited to a range $\Delta\nu$ that is much greater than the detector output frequency but small compared to kT/h, so that the power per unit optical frequency is constant. With this bandwidth $\Delta\nu$, the detector-current spectrum consists of frequencies from zero to $\Delta\nu$, because the highest frequency corresponds to beats between field fluctuations at the extremes of the optical bandwidth. The spectral density of the current is of the form

$$\overline{i_N^2}(f) \propto \int_{\nu}^{\nu+\Delta\nu} \overline{E^2}(\nu)\, \overline{E^2}(\nu+f)\, d\nu .$$

The resultant spectral density from Fig. 7.2 is

$$\overline{i_N^2}(f) = \overline{i_N^2}(0) \left(1 - \frac{f}{\Delta\nu}\right) .$$

Because the total mean-square current is the integral of $\overline{i_N^2}(f)$ over f, we obtain

$$\overline{i_N^2} = \int_0^{\Delta\nu} \overline{i_N^2}(f) df = \overline{i_N^2}(0) \frac{\Delta\nu}{2}; \quad \therefore\ \overline{i_N^2}(0) = \frac{2\overline{i_N^2}}{\Delta\nu} .$$

The final expression for the noise current becomes, from (7.2),

$$\overline{i_N^2} = 2e \frac{\eta e}{h\nu} B\bar{P} + 2 \left(\frac{\eta e}{h\nu}\right)^2 \frac{B}{\Delta\nu} \sum_k \bar{P}_k^2$$

for frequencies f much less than $\Delta\nu$.

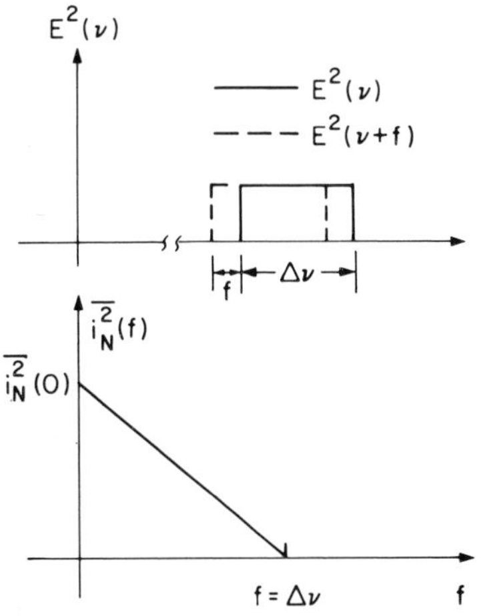

Fig. 7.2. Calculation of noise-current spectral density

For thermal radiation, the power per mode is

$$\overline{P_k} = \varepsilon \mathscr{f}_k h\nu \Delta\nu$$

where $\mathscr{f}$ is the Bose-Einstein occupancy probability. Therefore, we may rewrite the preceding expression as

$$\overline{i_N^2} = 2\frac{\eta e^2}{h\nu} B\bar{P} + 2\left(\frac{\eta e}{h\nu}\right)^2 \frac{B}{\Delta\nu} \cdot \sum_k \varepsilon^2 \mathscr{f}_k^2 (h\nu\Delta\nu)^2 .$$

For a small frequency range $\Delta\nu$, $\mathscr{f}_k$ is independent of k and the final expression for the noise current is

$$\overline{i_N^2} = 2\frac{\eta e^2}{h\nu} B\bar{P} + 2\eta^2 e^2 B\Delta\nu\, k\varepsilon^2 \mathscr{f}^2 ,$$

which we may write

$$\overline{i_N^2} = 2\frac{\eta e^2}{h\nu} B\bar{P} + 2\frac{\eta^2 e^2}{h\nu} \varepsilon\mathscr{f} B\bar{P} = 2\frac{\eta e^2}{h\nu} B\bar{P}\,(1 + \eta\varepsilon\mathscr{f}) . \qquad (7.3)$$

This is the correct expression for the background-induced noise current if the occupancy factor $\mathscr{f}$ is not small compared with unity. [It should be mentioned that there is inconsistency in the literature about the inclusion of η in this expression. This matter is discussed in detail by *Hanbury Brown* and *Twiss*, (1957a; Appendix II)]. In our derivations of D^* in Chapter 2, we ignored the power-fluctuation term, because $\mathscr{f}$ was much less than unity.

In the case of the thermal detector, we shall wish to know the *effective* fluctuation of the power incident upon the detector. In this case, we ask what the equivalent power fluctuation would be if the quantum efficiency were unity and the emissivity of the blackbody source were also unity. In this case we may rewrite the equation for the noise current as

$$\overline{i_N^2} = \frac{2e^2}{h\nu} B\bar{P}(1+f) = \left(\frac{e}{h\nu}\right)^2 [\Delta P^2(f)]_{\text{eff}} B,$$

where $\Delta P^2(f)_{\text{eff}}$ is the fluctuation of incident power that would produce the *measured* output-current noise if we ignored the noise associated with the photo-excitation process. We forget the measurement process and investigate the fluctuation of rate of arrival of "photons" by first solving for the effective power fluctuation,

$$[\Delta P^2(f)]_{\text{eff}} = 2h\nu\bar{P}(1+f). \tag{7.4}$$

As in the derivation of (7.3), the effective fluctuation of the total power becomes

$$(\overline{\Delta P^2})_{\text{eff}} = [\overline{\Delta P^2}(f)]_{\text{eff}} \frac{\Delta\nu}{2}$$

and

$$(\overline{\Delta P^2})_{\text{eff}} = h\nu\Delta\nu\bar{P}(1+f).$$

For a single mode, $P = P_k = f h\nu\Delta\nu$; therefore, with $r = P/h\nu$

$$\overline{\Delta r^2} = (\bar{r}\Delta\nu)(1+f) = (\Delta\nu)^2 f(1+f),$$

which states that the mean-square fluctuation of arrival rate for low occupancy, f, is proportional to the arrival rate $\bar{r}$, whereas for large occupancy the fluctuation is increased by the factor $(1+f)$. In quantum-mechanical terms, this is equivalent to saying that the mean-square fluctuation of occupancy of a Bose-Einsteim state is given by

$$\overline{\Delta f^2} = f(1+f),$$

a property derivable from quantum statistics. This so-called "bunching" of the photons is thus associated with the coherent interaction of the random electromagnetic waves; we have arrived at it semiclassically by associating the quantum process with the detector interaction with the field. When we treat the thermal detector, we should realize that energy absorption is a quantized process, in the sense that the detector material extracts energy from the wave only in discreate steps, equal to the photon energy. Thus, the power transferred from the radiation field to the detector material is defined by the *effective* power fluctuation, as in (7.4).

7.2 Sensitivity of the Ideal Thermal Detector

We now treat the thermal detector, which we define as an element whose temperature is a function of the incident radiation, which temperature can be monitored by some form of electrical measuring technique. We first derive the sensitivity in terms of the minimum measurable temperature change, as determined by the sensitivity of the element to radiation and by the inherent temperature fluctuations present in the element (*van der Ziel*, 1954).

First, consider a small detector element at temperature T in equilibrium with its chamber, also at temperature T. We assume that the element has a heat capacity $\mathscr{C}$, defined as $d\mathscr{E}/dT$, or the change of total energy with temperature in JK^{-1}, and a heat conductance $\mathscr{G}$ to its surroundings, given by dP/dT in WK^{-1}. The heat conductance is associated with the radiative heat transfer as well as with heat conduction by the electrical leads and supporting structures. For the ideal case considered here, we shall neglect the last terms and consider only the radiative loss.

We now ask what is the mean-square temperature fluctuation, because we know that the steady-state change of temperature per unit incident power is determined by $P/\mathscr{G}$. To find this quantity, we utilize the Boltzmann relationship, which tells us that the probability of the element having energy $\mathscr{E}_i$ is

$$p(\mathscr{E}_i) = A\mathrm{e}^{-\mathscr{E}_i/kT} \text{ with } \sum_i A\mathrm{e}^{-\mathscr{E}_i/kT} = 1\,.$$

The average energy of the body is simply

$$\mathscr{E} = \sum_i \mathscr{E}_i\, p(\mathscr{E}_i) = \sum_i A\mathscr{E}_i\,\mathrm{e}^{-\mathscr{E}_i/kT} = \frac{\sum_i \mathscr{E}_i\,\mathrm{e}^{-\mathscr{E}_i/kT}}{\sum_i \mathrm{e}^{-\mathscr{E}_i/kT}}\,.$$

From its definition, the heat capacity is

$$\frac{d\mathscr{E}}{dT} = \mathscr{C} = \frac{1}{kT^2}\left[\frac{\sum_i \mathscr{E}_i^2\mathrm{e}^{-\mathscr{E}_i/kT}}{\sum_i \mathrm{e}^{-\mathscr{E}_i/kT}} - \left(\frac{\sum_i \mathscr{E}_i\,\mathrm{e}^{-\mathscr{E}_i/kT}}{\sum \mathrm{e}^{-\mathscr{E}_i/kT}}\right)^2\right],$$

for the calculation of which we remember that the derivative of a sum is equal to the sum of the derivatives. The first term in the brackets is the *mean-square* energy, whereas the second term is the *square* of the *mean* energy. We thus obtain the mean-square energy fluctuation,

$$\frac{d\mathcal{E}}{dT}=\mathcal{C}=\frac{1}{kT^2}(\overline{\mathcal{E}^2}-\bar{\mathcal{E}}^2)=\frac{1}{kT^2}\overline{\Delta\mathcal{E}^2}.$$

Because $d\mathcal{E}/dT=\mathcal{C}$, the mean-square temperature fluctuation is

$$\overline{\Delta T^2}=\frac{\overline{\Delta\mathcal{E}^2}}{\mathcal{C}^2}=\frac{kT^2}{\mathcal{C}}.$$

For our purposes, we need the spectral density of the temperature fluctuation, that is, the mean-square fluctuation per unit frequency interval. Because the body has a conductance $\mathcal{G}$ and capacitance $\mathcal{C}$, it behaves as an RC circuit with time constant $\tau=\mathcal{C}/\mathcal{G}$. From basic network theory, we know that the power spectrum of such a body is of the form,

$$\overline{\Delta T^2}(f)=\frac{K}{1+\omega^2\tau^2}.$$

Thus

$$\overline{\Delta T^2}=\frac{kT^2}{\mathcal{C}}=\int_0^\infty\overline{\Delta T^2}(f)\,df=\frac{K}{2\pi}\int_0^\infty\frac{d\omega}{1+\omega^2\tau^2},$$

from which

$$\overline{\Delta T^2}=\frac{kT^2}{\mathcal{C}}=\frac{K}{2\pi\tau}\int_0^\infty\frac{dx}{1+x^2}=\frac{K}{2\pi\tau}\left(\frac{\pi}{2}\right)=\frac{K}{4\tau}=\frac{K\mathcal{G}}{4\mathcal{C}},$$

from which, after solving for K, we obtain

$$\overline{\Delta T^2}(f)=\frac{4kT^2}{\mathcal{G}}\left(\frac{1}{1+\omega^2\tau^2}\right).$$

For frequencies much less than $1/\tau$, the final temperature fluctuation, in terms of electrical bandwidth of the sensor, is

$$\overline{\Delta T^2}=\frac{4kT^2}{\mathcal{G}}B. \qquad (7.5)$$

In the presence of a radiation signal P_S, the change of temperature is

$$\overline{\Delta T_S}=P_S/\mathcal{G}.$$

For a signal-to-noise ratio of unity, the resultant *NEP* is found by equating the two temperature changes, which yields

$$NEP = \sqrt{4kT^2\mathscr{G}B}\,. \tag{7.6}$$

Thus, as we might expect, the smaller the conductance the higher the sensitivity, or the lower the *NEP*. We now must calculate $\mathscr{G}$. We do so by assuming unit emissivity (as we did for the signal-power absorption) and calculating the radiative loss, which is simply

$$\frac{dP}{dT} = \frac{d}{dT}(\sigma A T^4) = 4\sigma A T^3\,.$$

The final *NEP* for the ideal thermal detector is thus

$$NEP = \sqrt{4kT^2\mathscr{G}B} = 4\sqrt{A\sigma k T^5 B}\,, \tag{7.7}$$

which may be written more conveniently as

$$NEP = 4\sqrt{AB\cdot(\sigma T^4)(kT)}\,.$$

As a numerical example, let $T = 300$ K, for which the term σT^4 is 460 $\mathrm{Wm^{-2}}$ or 0.046 $\mathrm{Wcm^{-2}}$. Then D^* becomes

$$\begin{aligned} D^* &= \frac{\sqrt{AB}}{NEP} = \frac{1}{4\sqrt{(\sigma T^4)(kT)}} = \frac{1}{4\sqrt{(0.046)(4\times 10^{-21})}} \\ &= 1.8\times 10^{10}\ \mathrm{cm}\cdot\mathrm{Hz}^{1/2}\ \mathrm{W}^{-1}, \end{aligned} \tag{7.8}$$

which is within a factor of two of the minimum value of D^* for an ideal photodetector with a 180° FOV. Of course, the sensitivity is comparable only under this special condition and in the 10 to 15 μm region. In addition, the time constant of typical thermal detectors is such that bandwidths are generally limited to 10 to 1000 Hz.

Obviously, cooling of the detector and its chamber should improve the sensitivity. If T_0 is much less than the background temperature, the dominant fluctuation will be associated with fluctuations of the background power.

Assume a cooled detector of area A, and unity emissivity, exposed to a radiation field at temperature T, and FOV of half-angle $\theta/2$. The fluctuation of power for a small frequency interval $\Delta\nu$ is, from (7.4),

$$[\Delta P^2(f)]_{\mathrm{eff}} = 2h\nu\bar{P}(1+\mathscr{f})\,.$$

Integrating over all frequency ν thus yields

$$\overline{\Delta P^2} = \int_{\nu=0}^{\infty} 2h\nu B(1 + f_\nu)\, dP_\nu \,.$$

From (2.21), the power striking a surface over full cone-angle θ is

$$dP_\nu = A \sin^2 \frac{\theta}{2} \frac{2\pi h\nu^3 d\nu}{c^2(e^{h\nu/kT} - 1)} = A \sin^2 \frac{\theta}{2} \frac{2\pi h\nu^3 f d\nu}{c^2} \,.$$

The quantity $f(1 + f)$ may be written

$$f(1 + f) = \frac{e^{h\nu/kT}}{(e^{h\nu/kT} - 1)^2}$$

and the final expression for the mean-square power fluctuation is

$$\Delta P^2 = 2AB \sin^2 \frac{\theta}{2} \int_0^{\infty} \frac{2\pi\nu^2}{c^2} \frac{(h\nu)^2\, e^{h\nu/kT}\, d\nu}{(e^{h\nu/kT} - 1)^2} \,.$$

This integral may be written, after reduction of the integrand, as

$$\overline{\Delta P^2} = 2AB \sin^2 \left(\frac{\theta}{2}\right) \left[\frac{2\pi(kT)^5}{c^2h^3}\right] \int_0^{\infty} \frac{x^4 e^x dx}{(e^x - 1)^2} = 2AB \sin^2 \frac{\theta}{2} \left(\frac{2\pi(kT)^5}{c^2h^3}\right) \left(\frac{4\pi^4}{15}\right) .$$

The value of σ, the Boltzmann constant, is

$$\sigma = \frac{2\pi^5 k^4}{15c^2h^3} \,.$$

The final result is

$$\overline{\Delta P^2} = 2AB \sin^2 \frac{\theta}{2} \cdot 4\sigma k T^5 = 8 \sin^2 \frac{\theta}{2} AB\sigma k T^5 \,.$$

For unit signal-to-noise ratio,

$$(NEP)^2 = 8 \sin^2 \frac{\theta}{2} AB\sigma k T^5 \,.$$

Note that this expression, for $\theta = 180°$, is just one-half that obtained for the detector limited by the thermal bath. The reason is that the temperature fluctuations of the detector element are determined by the fluctuations of the *emitted* power and the *absorbed* power. Because these fluctuations are independent, they add together and give *twice* the value, if the detector element is in equilibrium with the incident-radiation field. We have thus made another direct connection between Boltzmann statistics and the statistics of the Bose-Einstein radiation field.

Finally, we consider the case of a thermal detector at low temperature, which is exposed only to radiation above a certain optical cutoff frequency ν_c, where $h\nu_c \gg kT$. In that case,

$$\overline{\Delta P^2} = \int_{\nu_c}^{\infty} 2\,h\nu B(1+\mathscr{f})\,dP_\nu = \int_{\nu_c}^{\infty} 2h\nu B dP_\nu\,,$$

because $\mathscr{f}$ is much less than 1. The integral then becomes

$$\overline{\Delta P^2} = \int_{\nu_c}^{\infty} 2h\nu B\,dP_\nu = 2h\nu B \cdot P_B\,,$$

where the latter equality holds to the extent that $h\nu$ varies only slightly over the region where the $\exp(-h\nu/kT)$ term in the power flux falls off rapidly. Thus, the ideal thermal detector with a cooled filter approaches the sensitivity of a perfect photon counter.

We now consider practical forms of thermal detectors and treat the theory for two of them in greater detail. The thermocouple or thermopile was the earliest detector; it depends upon the thermoelectric emf produced between two metals of different thermoelectric power in the presence of a thermal gradient. This device is simple, but the small value of the thermoelectric coefficient limits its sensitivity, because of amplifier noise. A second and much more sensitive device is the Golay cell, which consists of an absorbing thin membrane immersed in a heat-conducting gas. Incident radiation heats the membrane, which in turn heats the gas; the resultant pressure change is sensed by a second light-reflecting membrane which is distorted in such a way as to deflect a reflected sampling optical beam, which deflection is sensed by a standard visible detector. The low thermal capacity and conductance of the thermal-absorbing membrane make this device extremely sensitive, although the frequency response is limited by the thermal time constant of the system. The theory of this device was described by *Golay* (1949). The bolometer and the pyroelectric detector will next be treated in detail.

7.3 Bolometers

Bolometers are particularly effective devices when cooled below ambient, since their response can approach that of the ideal thermal detector discussed in the previous section. The bolometer (sometimes called a thermistor) is a detector element whose electrical resistance is a strong function of temperature. Semiconductors are the most favored materials although some devices depend upon the superconducting transition to obtain a rapid change of resistance with temperature. The appropriate coefficient for defining the response of these detectors is the coefficient $b = (1/R)dR/dT$, or the fractional change in resistance per degree Kelvin. We now assume that such a device is driven by a constant current and is enclosed in a chamber at temperature T_0. Since the current through the resistor heats it slightly, we define the temperature of the detector element as T, and assume that $(T - T_0) \ll T_0$. The signal voltage then becomes (*van der Ziel*, 1954)

$$V_S = I\Delta R = IbR\Delta T = IbR\frac{P_S}{\mathscr{G}},$$

and assuming an amplifier with effective input noise temperature T_N, the mean-square noise voltage is then

$$\overline{V_N^2} = 4kT_N RB + I^2b^2R^2\,\overline{\Delta T^2},$$

where the last term is the mean-square voltage fluctuation associated with the temperature fluctuation of the body. From (7.5), a body in thermal equilibrium with a temperature bath T has a temperature fluctuation given by

$$\overline{\Delta T^2} = \frac{4kT^2}{\mathscr{G}}B;$$

we showed in Section 7.2 that one-half of this fluctuation was due to fluctuating *emission* of radiation and the other half due to fluctuating *absorption*. It follows that the mean-square temperature fluctuation in the bolometer case is

$$\overline{\Delta T^2} = \left(\frac{2kT_0^2}{\mathscr{G}} + \frac{2kT^2}{\mathscr{G}}\right)B = \frac{4kT_0^2}{\mathscr{G}}\left(1 + \frac{\Delta T}{T_0}\right)B,$$

where we have used the fact that ΔT, the temperature differential due to the current heating, is small compared with T_0. But from the definition of the heat conductance $\mathscr{G}$, we may rewrite this term as

$$\overline{\Delta T^2} = \frac{4kT_0^2}{\mathscr{G}}\left(1 + \frac{\Delta T}{T_0}\right) = \frac{4kT_0^2}{\mathscr{G}}\left(1 + \frac{I^2R}{\mathscr{G}T_0}\right).$$

The resultant voltage fluctuation is

$$\overline{V_N^2} = 4kT_N RB + \frac{4I^2b^2R^2kT_0^2}{\mathscr{G}}\left(1 + \frac{I^2R}{\mathscr{G}T_0}\right)$$

and the signal-to-noise ratio is

$$\left(\frac{S}{N}\right)_{\mathrm{P}} = \frac{V_{\mathrm{S}}^2}{V_{\mathrm{N}}^2} = \frac{I^2 b^2 R^2 P_{\mathrm{S}}^2/\mathscr{G}^2}{\dfrac{4I^2b^2R^2kT_0^2}{\mathscr{G}}\left(1 + \dfrac{I^2R}{\mathscr{G}T_0}\right) + 4kT_{\mathrm{N}}RB}\,.$$

For $S/N = 1$, we find the square of the *NEP* to be

$$(NEP)^2 = 4kT_0^2\mathscr{G}B + 4kT_0I^2RB + \frac{4kT_{\mathrm{N}}\mathscr{G}^2B}{I^2b^2R}\,.$$

The third term indicates a decreasing *NEP* with current, associated with the increased signal response, whereas the second term increases the *NEP* because it is associated with an increase of detector temperature and resultant increased temperature fluctuation, which produce more noise. Differentiating the last two terms with respect to current, we find the minimum *NEP* for

$$I^2 = \sqrt{\frac{T_{\mathrm{N}}}{T_0}}\frac{\mathscr{G}}{|b|R}\,.$$

For this optimum current, the square of the *NEP* is

$$(NEP)^2 = 4kT_0^2\mathscr{G}B\left(1 + \frac{2\sqrt{T_{\mathrm{N}}/T_0}}{|b|T_0}\right),$$

which is the same as the expression for the ideal thermal detector (7.6), except for the factor in parentheses. As an example of this term, consider a semiconductor bolometer that operates at approximately 4K. The coefficient b is

$$b = \frac{1}{R}\frac{dR}{dT}\,;\; R \approx R_0\left(\frac{T_0}{T}\right)^n \;\therefore\; \mathrm{b} = -\frac{n}{T_0} \approx \frac{-4}{T_0}\,,$$

where the variation of R has been determined empirically. The value of $(|b|\,T_0)$ is thus 4 and the *NEP* is increased by the square root of the term $[1 + 0.5\sqrt{T_{\mathrm{N}}T_0}]$. The D^* for this detector, assuming only radiative conductance, is thus *greater* than the ideal value of (7.8) by a factor of $(300/4)^{5/2}$ but reduced by $[1 + (300/4)^{1/2}/2]$ for $T_{\mathrm{N}} = 300\mathrm{K}$. This results in a D^* approximately $(75)^2$ times that of the ideal room-temperature detector or approximately 10^{14}, a very respectable value. Actually, the low thermal conductance at 4K yields extremely long time constants; typical detectors are operated with a conduction path that decreases the detectivity but also the response time. The semiconductor bolometer is treated in great detail and with more rigor by *Low* (1961).

7.4 The Pyroelectric Detector

For room-temperature operation, the pyroelectric detector offers the best performance, both because of its high sensitivity and its frequency response.

This device is based upon a temperature-sensitive ferroelectric material such as triglycine sulfate, which has a net charge transfer proportional to temperature. The current from such a detector is proportional to the rate of change of temperature and is given by

$$i = A\left(\frac{d\mathscr{P}}{dT}\right) \cdot \frac{dT}{dt} = AK_{\mathrm{p}} \cdot \frac{dT}{dt},$$

where $K_{\mathrm{p}} = (d\mathscr{P}/dT)$ is the rate of change of the polarization with temperature. The signal current is thus given by

$$i_{\mathrm{S}} = K_{\mathrm{p}} A\omega \Delta T = K_{\mathrm{p}} A\omega \frac{P_{\mathrm{S}}}{\mathscr{G}},$$

and the mean-square noise current by

$$\overline{i_{\mathrm{N}}^2} = \frac{4kT_{\mathrm{N}}B}{R} + K_{\mathrm{p}}^2 A^2 \omega^2 \overline{\Delta T^2} = \frac{4kT_{\mathrm{N}}B}{R} + 4K_{\mathrm{p}}^2 A^2 \omega^2 \frac{kT^2}{\mathscr{G}} B,$$

which result in a signal-to-noise ratio of

$$\left(\frac{S}{N}\right)_{\mathrm{P}} = \frac{\overline{i_{\mathrm{S}}^2}}{\overline{i_{\mathrm{N}}^2}} = \frac{P_{\mathrm{S}}^2}{4kT^2\mathscr{G}B + \dfrac{4kT_{\mathrm{N}}B\mathscr{G}^2}{RK_{\mathrm{p}}^2A^2\omega^2}},$$

and an *NEP* of

$$(NEP)^2 = 4kT^2\mathscr{G}B\left(1 + \frac{T_{\mathrm{N}}}{T^2} \cdot \frac{\mathscr{G}}{K_{\mathrm{p}}^2A^2\omega^2R}\right).$$

Of particular interest is the decrease of *NEP* with increased frequency. Actually, the thermal time constant limits the response at a low frequency, of the order of 1 to 10 Hz; however, the response and signal-to-noise ratio then remain constant out to the *RC* cutoff frequency of the electrical circuits. Thus high-frequency response may be obtained at the sacrifice of sensitivity. Taking into account the temperature response in terms of the thermal capacitance and conductance, the value of *NEP* may be rewritten as

$$(NEP)^2 = 4kT^2\mathscr{G}B\left(1 + \frac{T_{\mathrm{N}}\mathscr{C}^2}{T^2K_{\mathrm{p}}^2A^2R\mathscr{G}}\right) \tag{7.9}$$

for ω much greater than $\mathscr{G}/\mathscr{C}$. At $\omega = \mathscr{G}/\mathscr{C}$, (7.9) reduces to the low-frequency value originally derived. In any event, both the signal power and the temperature-induced noise power behave as

$$\overline{i_S^2} \propto \overline{i_N^2} \propto \frac{\omega^2}{1+\omega^2\tau^2} \longrightarrow \frac{1}{\tau^2} = \frac{\mathscr{G}^2}{\mathscr{C}^2}$$

and thus remain constant in the high-frequency range up to the frequency at which the *electrical* capacitance limits the response.

As an example of proelectric detection, we consider triglycine sulphate, the most effective of the pyroelectric materials. The important physical constants of this material at room temperature are:

$$K_p = 2 \times 10^{-8} \quad \mathrm{C\,cm^{-2}\,K^{-1}}$$

$$c = 1.64\ \mathrm{J\,cm^{-3}\,K^{-1}}$$

$$\kappa = 25.$$

Taking a detector area 1 mm^2 or 10^{-2} cm^2 and a thickness 10 μm or 10^{-3} cm, we calculate the *electrical* capacitance as

$$C = \frac{k\varepsilon_0 A}{t} = \frac{25(8.85 \times 10^{-14})(10^{-2})}{10^{-3}} = 22 \times 10^{-12} = 22\ \mathrm{pF}\,.$$

If we wish an electrical cutoff frequency of 1 kHz, this yields the required load resistance,

$$R = \frac{1}{2\pi f_c C_c} = 7 \times 10^6\ \Omega\,.$$

The thermal capacitance and conductance are given by

$$\mathscr{C} = cAt = 1.64 \times 10^{-5}\ \mathrm{JK^{-1}}$$

$$\mathscr{G} = 4\sigma AT^3 = \frac{4(\sigma T^4)A}{T} = \frac{4(0.046)\,10^{-2}}{300} = 6 \times 10^{-6}\ \mathrm{WK^{-4}}\,.$$

The thickness t should be as small as possible, because, for a given frequency response, R is proportional to t, but $\mathscr{C}^2$ in the excess noise factor increases as t^2. Assuming $T_N = T = 300$ K, we obtain for the second term in parentheses in (7.9)

$$\frac{T_N\mathscr{C}^2}{T^2K_p^2A^2R\mathscr{G}} = \frac{(300)(1.64 \times 10^{-5})^2}{(300)^2(2 \times 10^{-3})^2(10^{-4})(7 \times 10^6)(6 \times 10^{-6})} = 5 \times 10^5\,.$$

Thus, the value of $(NEP)^2$ is increased by this ratio over that of an ideal thermal detector. Therefore, we may write

$$NEP = \sqrt{5 \times 10^5}\, NEP_{\text{ideal}} = \frac{\sqrt{5 \times 10^5}\,\sqrt{AB}}{1.8 \times 10^{10}} = 3.5 \times 10^{-9}\,\sqrt{B}\ \text{W}$$

or for full baseband response to 1 kHz, the *NEP* is 10^{-7} W. Because the excess noise factor is proportional to $B^{1/2}$ through the resistance *R*, the *NEP* for full baseband detection is proportional to *B*. Thus, for 10^9 MHz bandwidth, equivalent to nanosecond rise times, the *NEP* becomes 0.1 W. This is of course not an impressive sensitivity, but turns out to be extremely useful in laboratory measurements for which the only alternative in the long-wavelength infrared region would be a cyrogenically cooled semiconductor device. A through discussion of pyroelectric detectors may be found in *Putley* (1970, 1977).

7.5 Heterodyne Detection with Thermal Detectors

Although of somewhat limited application, heterodyne detection is perfectly feasible with a thermal detector, but with extremely limited frequency response for sensitivity that approaches the ideal value of $P_S/h\nu B$. Consider a general thermal detector upon which is incident signal and local-oscillator power, P_S and P_{LO}. We define the current (or voltage) in the output circuit in terms of a responsivity r as $i = rP$. From the treatment of Chapter 3, the heterodyne i.f. current is

$$i_{\text{i.f.}}^2 = 2i_S i_{LO} = 2r^2 P_S P_{LO}\,,$$

subject of course to the restriction that the signal and local-oscillator powers are in the same spatial mode of free space, that is, that the mixing efficiency is unity. In the absence of local-oscillator radiation, the effective mean-square power fluctuation at the detector is

$$(\overline{\Delta P^2}) = (NEP)^2\,,$$

whereas the local-oscillator power adds a term produced by its effective mean-square fluctuation *as sensed by the detector*. This is, from (7.4),

$$(\overline{\Delta P^2})_{LO} = 2h\nu B\bar{P}(1 + \ell) \rightarrow 2h\nu B\bar{P}_{LO}\,,$$

because we must set $\ell = 0$, because $(\overline{\Delta P^2})_{LO}$ represents the *fluctuation* of power associated with the classical Gaussian field statistics. Because P_{LO} is obtained from a laser, which has *constant* amplitude power output, there is no such fluctuation term. It is interesting to note this apparent dichotomy in the significance of f, the occupation probability of an electromagnetic mode. Specifically, the occupancy of a mode associated with laser radiation is astronomically high—$P/h\nu\Delta\nu$, where $\Delta\nu$ is the spectral linewidth of the radiation. Yet, the noise or effective power fluctuation is found by setting $f = 0$ in the last expression. Actually, because another factor of f is buried in $\bar{P}$, the effective laser noise is proportional to f rather than to $f(1 + f)$, as is the case with thermal radiation, provided that the laser power is constant, as is nominally the case, because it is a saturated oscillator.

Returning now to the signal and noise currents, we may write

$$i_N^2 = r^2(\overline{\Delta P^2})_{\text{eff}} = r^2\,[(NEP)^2 + 2h\nu B\bar{P}]\,.$$

Solving for the signal-to-noise ratio, we obtain

$$\left(\frac{S}{N}\right)_P = \frac{2P_{LO}P_S}{(NEP)^2 + 2h\nu B\bar{P}_{LO}}\,,$$

which approaches ideal heterodyne detection, provided that

$$P_{LO} \gg \frac{(NEP)^2}{2h\nu B}\,.$$

For an ideal thermal detector, the requirement becomes

$$P_{LO} \gg \frac{16A\sigma kT^5B}{2h\nu B} = \frac{8A\sigma kT^5}{h\nu}\,.$$

Because the background is $P_B = A\sigma T^4$ for a detector in equilibrium with its surroundings, we may write

$$P_{LO} \gg \frac{8kT}{h\nu}P_B\,.$$

By comparison, the ideal photodetector requires a local-oscillator power much greater than the single term P_B. Finally, we take a real thermal detector and ask for the required local-oscillator power. As an example, consider the pyroelectric detector discussed previously and assume operation at 10 µm. This case, for a bandwidth of 1 kHz, and the calculated NEP of 10^{-7} W, results in a power requirement of

$$P_{LO} \gg \frac{(NEP)^2}{2h\nu B} = \frac{10^{-14}}{2(0.12)\,(1.6 \times 10^{-19})\,(10^3)} = 260\ \text{W}\,.$$

This power is obviously impractical, in terms of heating of the detector. We note that the power requirement is proportional to bandwidth because the *NEP* also is proportional to bandwidth. Thus, except for bandwidths much less than 1 Hz, the device does not yield ideal heterodyne detection. Obviously, a cooled thermal detector would offer more promise, and operation at a much shorter wavelength would reduce the power requirement. In any event, heterodyne detection with reduced sensitivity is perfectly feasible, with sensitivity

$$\left(\frac{S}{N}\right)_{\mathrm{P}} = \frac{2P_{\mathrm{S}}P_{\mathrm{LO}}}{(NEP)^2}\,.$$

In systems or laboratory experiments in which sensitivity is not a critical factor, the use of a thermal detector for heterodyne measurements has many advantages, especially at long wavelengths where photon detectors require complicated cooling apparatus.

Problems

7.1 An ideal thermal detector is operated at a temperature T_{d}, in a cavity at the same temperature. Incident upon the detector, whose area is A, are local oscillator power P_{LO} and signal power P_{S}. Assume that the background is negligible, and find the *heterodyne* signal-to-noise ratio, neglecting the noise in the sensing element or amplifier.

7.2 If the element is of area 1 mm^2, find the required local oscillator power for LO noise equal to thermal noise at the temperature $T_{\mathrm{d}} = 77$ K and at 300 K.

7.3 If the heat capacity of the detector is 10^{-7} JK^{-1}, find the time constant (thermal) of the detector at 300 K.

7.4 The above value obviously limits the bandwidth for heterodyne detection. Now assume a heat-conduction path to the detector such that $\tau = 10^{-6}$ s corresponding to a bandwidth of about 1 MHz. What is the required shunting thermal conductance $\mathcal{G}$?

7.5 Find the *NEP* (incoherent) with this shunt conductance.

7.6 What is the new required local oscillator power at $T = 300$ K?

8. Laser Preamplification

One possible way to improve the sensitivity of an optical or infrared detector is to use a laser as a preamplifier for the incoming radiation. This might be especially advantageous in the infrared region, where amplifier noise limits the ultimate sensitivity in the incoherent detection process. We shall here give a general treatment of the noise limitations in a laser preamplifier and show that the ultimate signal-to-noise performance is almost identical to that of a heterodyne detector.

We start with the configuration of Fig. 8.1, in which a detector receives the amplified signal and noise produced in a lasing medium, when the signal power is incident on the left of the medium. For the moment, we shall specify that the *detector* is coherent, that is, a heterodyne detector that couples to only *one* mode of free space. This assumption is essential to calculate the spontaneous emission noise of the laser medium in a simple manner. Our treatment is similar to that of *Yariv* (1975), except that use of the single-spatial-mode concept at the outset simplifies the treatment. Let us describe the medium in terms of the two quantities n_1 and n_2, the occupancies of the lower and upper states of the laser transition. To calculate the spontaneous-emission noise, we first assume that the ensemble is at thermal equilibrium and in an absorbing condition. In this case, from the discussions in Sections 1.3 and 3.4, we know that the noise power emitted per mode of free space over a small distance dx is given by

$$dP = (\alpha_l \, dx) \frac{h\nu \Delta\nu}{(e^{h\nu/kT} - 1)}, \qquad \Delta\nu \ll \frac{kT}{h},$$

where α_l is the absorption coefficient and thus $\alpha_l dx$ is the emissivity. We next note that the absorption coefficient is given by

$$\alpha_l = C(n_1 - n_2),$$

where the constant C contains the B coefficient of Section 1.4, which is a measure of the induced-transition efficiency. The factor $\exp(h\nu/kT)$ in the emission noise equation is simply the ratio of n_1 to n_2, if we assume a Boltzman occupancy law at thermal equilibrium, so we may write

$$dP = C(n_1 - n_2) \, dx \frac{h\nu \Delta\nu}{\left(\frac{n_1}{n_2} - 1\right)} = Cn_2 h\nu \Delta\nu dx,$$

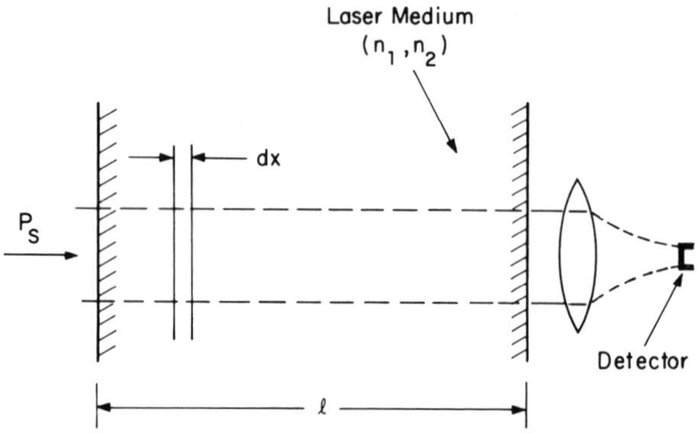

Fig. 8.1. Laser medium with detector

which states that the spontaneous-emission power *emitted into a single mode* is proportional to n_2, as we would have expected. The reason why we use this apparent roundabout way to derive the spontaneous emission is that the normal form of the emission term, which involves the A coefficient, describes the total radiation in all directions, rather than into a single mode. We now initiate the laser action, which results in an inversion of the level populations and even a change of total occupancy. Under these conditions, the spontaneous-emission power per unit distance dx remains the same but with a new value of n_2 because we are assuming low occupancy through use of Boltzmann statistics. (We should interject that this treatment is equally applicable to levels governed by Fermi statistics, just as we found in Problem 1.5 for the thermal-radiation-field interaction with a two-level system). We now assume that the laser medium has a *gain* coefficient α_g, given by

$$\alpha_g = C(n_2 - n_1) \tag{8.1}$$

because the constant C is a measure of the induced-transition rate vs incident power. The equation for the buildup of spontaneous emission power may be written

$$dP_n = \alpha_g P_n(x)dx + Cn_2\, h\nu\Delta\nu dx\,,$$

where the first term on the right results from the amplification of the medium, and the second term is the added spontaneous emission power. This equation is solved by integrating from 0 to l, the length of the laser medium, with the boundary condition that $P_n = 0$ at $x = 0$. The result is

$$\int_0^{P_n(\mathrm{OUT})} \frac{dP_n}{\alpha_g P_n + Cn_2 h\nu\Delta\nu} = \int_0^l dx = l\,, \tag{8.2}$$

which yields

$$\frac{1}{\alpha_g} \ln \left[\frac{\alpha_g P_{n(OUT)} + C n_2 h\nu \Delta\nu}{C n_2 h\nu \Delta\nu} \right] = l$$

$$P_{n(OUT)} = (e^{\alpha_g l} - 1) \frac{C n_2 h\nu \Delta\nu}{\alpha_g} .$$

The term $\exp(\alpha_g l)$ is simply the power gain of the amplifier. Substituting for α_g from (8.1) leads to the final result,

$$P_{n(OUT)} = (G - 1) \frac{n_2}{(n_2 - n_1)} \cdot h\nu \Delta\nu . \tag{8.3}$$

This power, $P_{n(OUT)}$, is the total spontaneous-emission power produced in the amplifier, both from direct emission and also from subsequent amplification. If this power were constant, there would be no noise contribution in the detector except for the associated photon or "shot" noise. Actually, the phase and amplitude of the electric field produced by the emitting laser levels are random, exactly the same as are those of the thermal noise described in Section 7.1; the mean-square fluctuation of the laser power is given by the square of the mean as in (7.1). The root-mean-square power fluctuation is thus also given by $P_{n(OUT)}$. For n_2 much greater than n_1, or complete inversion, the output signal-to-noise ratio is

$$\left(\frac{S}{N}\right)_P = \frac{G P_S}{(G - 1) h\nu \Delta\nu} = \frac{P_S}{h\nu \Delta\nu} \text{ for } G \gg 1 , \tag{8.4}$$

with the very important proviso that this is the signal-to-noise ratio for amplification in a *single* mode of free space. (The *Yariv* treatment also includes the degeneracy of the initial and final states in the final noise expression; the reader is referred to this more general form). We therefore see that, for heterodyne detection, the laser preamplifier offers ideal detection, assuming complete inversion, because the detector responds to only one mode of free space; specifically, the $A\Omega$ product is equal to λ^2 for the detector or for the receiving aperture. In addition, because the heterodyne detector is a frequency converter, the spectral width $\Delta\nu$ is equivalent to an rf bandwidth B.

The foregoing arguments apply to a coherent or heterodyne detector; we now ask about the behavior when an incoherent detector is used at the output of the laser preamplifier. The resultant signal-to-noise ratio at the input of the detector then becomes

$$\left(\frac{S}{N}\right)_P = \frac{P_S}{N h\nu \Delta\nu} ,$$

where N is the total number of spatial modes intercepted by the device, including the double degeneracy associated with two possible polarizations. Thus, if we wish to obtain the optimum signal-to-noise ratio, we should limit the detector size to an area that satisfies $A\Omega = \lambda^2$, where Ω is the solid angle subtended by the field of view of the detector. In addition, we should insert a polarizer after the output of the laser. Note, however, that the field of view of the detector or receiver is limited to a single mode, just as in heterodyne detection. We note also another serious limitation, that the noise is proportional to *spectral* bandwidth $\Delta\nu$, that is, to the width of the optical-frequency spectrum of the received radiation, which, if there are no optical-frequency filters beyond the laser amplifier, is approximately equal to the laser linewidth. Let us insert such a filter and assume a spectral width of the *signal* power determined by the pulsewidth. In this case, $(S/N)_P$ will indeed approach the ideal value $P_S/h\nu B$, *provided* that the transmitter or source has spectral purity comparable to the inverse pulsewidth. In addition, we require an optical filter, perhaps a Fabry-Perot interferometer, that has the requisite narrow bandwidth to match the information bandwidth of the signal. In conclusion, the use of the laser as a preamplifier has serious restrictions when applied to an incoherent detection system. Although applicable to coherent detection, the expected sensitivity improvement over a heterodyne system is only $1/\eta$, where η is the quantum efficiency of the competing detector. Otherwise, the behavior is identical. We shall discuss the relative merits of incoherent and coherent detection in Chapter 11, in connection with specific system applications.

Problems

8.1 Derive the output noise power from a laser preamplifier, assuming Fermi-Dirac statistics for the levels n_1 and n_2. (See Problem 1.5).

8.2 A heterodyne detector is used at the output of a laser preamplifier. Assume that the laser gain G is much greater than unity, and that $n_2 \gg n_1$.

a) Assume that the detector intercepts N spatial modes and find the mean-square-noise current due to incoherent detection.

b) What is the total mean-square i.f. noise current with a local-oscillator power P_{LO}.

c) The incoherently detected noise occurs at baseband and the heterodyne converted noise at the i.f. frequency. If these bands overlap, show that

$$P_{LO} \gg \frac{NGh\nu B}{2}$$

will assure a final output $(S/N)_P$ of $P_S/h\nu B$.

9. The Effects of Atmospheric Turbulence

As discussed in Chapters 1 and 2, the absorption and reemission of radiation in the atmosphere affect the signal level at the detector and also limit the sensitivity, in terms of the background power. In addition, aerosol and molecular scattering are particularly important background sources in the visible region of the spectrum, because of the sun's radiation. Detailed data on absorption and scattering may be found in *Wolfe* (1965) and in the RCA *Electro-Optics Handbook*, hereinafter referred to as EOH (1974). We shall treat some of these effects in Chapter 11, but here shall consider another atmospheric property, turbulence, which has a profound effect upon heterodyne as well as incoherent-detection efficiency. The literature on turbulence is voluminous, and has recently been reviewed by *Fante* (1975). We shall here give a simple one-dimensional treatment based upon the work of *Fried* (1967a). In the first section, we find the limitation on effective aperture area in a heterodyne system and in the second, the minimum field of view allowed for an incoherent detector.

9.1 Heterodyne-Detection Limitations

We start by asking for the effective quantum efficiency of a heterodyne system in the presence of an incident plane wave whose wavefront has been distorted by passage through the atmosphere. Although we shall consider only phase distortions and a one-dimensional receiver aperture, *Fried* (1967a) treats the amplitude fluctuations as well as the circular-aperture case in great detail, but with much more complexity. We start with the function

$$D_\phi(r) = \overline{[\phi(x) - \phi(x')]^2}\,,$$

which is defined as the phase structure function and indicates the mean-square phase fluctuation between two points on the wavefront as a function of the separation of the points, $r = x - x'$. As detailed in *Fried*'s paper this may be obtained from turbulence theory and is given by

$$D_\phi(r) = \frac{k_1}{\lambda^2}\, r^{5/3} \int C_n^2\, dl = \frac{k_2 r^{5/3}}{\lambda^2}\,,$$

where the integral of C_n^2 over the path length is called the optical defect and C_n^2 is a measure of the local turbulence along the path. (There is a similar expression for the mean-square fluctuation of the logarithm of the amplitude, which is treated by *Fried*). The $r^{5/3}$ variation is peculiar to turbulence theory, whereas the λ^2 dependence comes from the fact that the phase shift is proportional to distance over wavelength; therefore, a given path-length perturbation results in a mean-square phase shift inversely proportional to the square of the wavelength. Returning to the heterodyne case, we write the fractional current efficiency as

$$\alpha(t) = \frac{i_T(t)}{i_0} = \frac{\int E_{ST}(x,t)\, E^*_{LO}(x)\, dx}{\int E_{SO}(x)\, E^*_{LO}(x)\, dx} ,$$

with

$$E_{ST}(x,t) = E_{SO} e^{i\phi(x,t)} ,$$

where E_{SO} is the unperturbed signal wave and E_{ST} is the distorted wave produced by the turbulence. Assuming a one-dimensional receiver that extends from $x = -D/2$ to $x = +D/2$, and that the fields, E_{SO} and E_{LO} are plane waves of constant amplitude at the aperture (note that we have invoked the mixing theorem at this point), we obtain for the current ratio

$$\alpha(t) = \int_{-D/2}^{+D/2} e^{i\phi(x,t)}\, dx ,$$

and must now determine the time average of this quantity. We average over times long compared with the characteristic time constant of the turbulence fluctuations, which is the order of 10^{-2} to 1 s. Before averaging, we rewrite the integral as

$$\int_{-D/2}^{+D/2} e^{i\phi(x,t)}\, dx = \int_{-D/2}^{+D/2} e^{i[\phi(x,t)-\phi(0,t)]}\, e^{i\phi(0,t)}\, dx .$$

The second exponential is simply a measure of the phase fluctuation at $x = 0$ and is independent of x. Its contribution is to produce a phase fluctuation of the detected current, so that we may ignore it in terms of the magnitude of the mixing efficiency. The final expression for the mean current ratio is

$$\langle \alpha \rangle = \frac{2}{D} \int_{-D/2}^{+D/2} \langle e^{i[\phi(x,t)-\phi(0,t)]} \rangle\, dx ,$$

where we have taken the average under the integral sign, because we are averaging over time and the operations are commutative. We now expand the exponential in a Taylor series and assume that the mean-square phase fluctuation is unity or less,

$$\langle\alpha\rangle = \frac{1}{D}\int_{-D/2}^{+D/2}\left\langle\left\{1 + \mathrm{i}[\phi(x,t) - \phi(0,t)] - \frac{1}{2}[\phi(x,t) - \phi(0,t)]^2 \ldots + \ldots\right\}\right\rangle dx\,.$$

The odd terms in the phase average to zero; the quartic term, if we had included it, would be divided by 4!, so that we may ignore it also. The final result is

$$\langle\alpha\rangle = \frac{1}{D}\int_{-D/2}^{+D/2}\left\langle\left\{1 - \frac{1}{2}[\phi(x,t) - \phi(0,t)]^2\right\}\right\rangle dx\,,$$

which from the definition of $D_\phi(r)$ leads to

$$\langle\alpha\rangle = \frac{2}{D}\int_0^{D/2}\left[1 - \frac{D_\phi(x)}{2}\right] dx\,,$$

where we integrate only from 0 to $+D/2$, because the r variable is the *magnitude* of $(x - x')$ and, by symmetry, the contributions from each side are equal. The final efficiency is

$$\langle\alpha\rangle = \frac{2}{D}\int_0^{D/2}\left[1 - \frac{1}{2}\frac{k_2}{\lambda^2}x^{5/3}\right] dx = 2/D\left[x - \frac{3}{16}\frac{k_2}{\lambda^2}x^{8/3}\right]_0^{D/2}\,,$$

which after a bit of manipulation yields

$$\langle\alpha\rangle = 1 - \frac{3}{16\cdot 2^{5/3}}\frac{k_2}{\lambda^2}D^{5/3}\,.$$

We now ask at what value of D does the mean current ratio fall to 50%; we obtain

$$\frac{r_0^{\,5/3}}{\lambda^2} = \frac{32\cdot 2^{5/3}}{3k_2}\,,$$

where r_0, the coherence diameter, is the value of D that satisfies this condition. This value is *not* the same as that found in the literature but will suit our purposes, for this simple derivation. We have simplified the mathematics by calculating a mean current rather than mean-square current ratio; however the form of the result is still valid. Rewriting the coherence diameter in terms of the turbulence parameter k_2 and the wavelength, we obtain

$$r_0 = \left(\frac{16\cdot 2^{5/3}}{3k_2}\right)^{3/5}\cdot\lambda^{6/5} \tag{9.1}$$

A typical value of r_0 for good astronomical seeing is 2 m for a wavelength of 10 µm, or by the above expression, about 12 cm for a wavelength of 1 µm (*Gilmartin* and *Holtz*, 1974). Although not obvious from our treatment, *Fried* shows that the collection efficiency $D^2\alpha^2$, or the signal power for a circular aperture saturates, for $D \gg r_0$, at a value approximately 2 times that at $D = r_0$, where at $D = r_0$; $\langle\alpha^2\rangle = 1/2$. Thus, increasing the aperture much beyond a diameter r_0 results in no appreciable increase of detected signal. We thus have concluded that heterodyne-detection systems are limited to apertures determined by the turbulence of the intervening medium.

9.2 Incoherent-Detection Limitations

We now ask about incoherent systems, by examining the fluctuations of angle of arrival of the wavefront, because, even in the incoherent case, this will determine the required field of view of the telescope system, and consequently the size of the detector. We proceed by calculating the mean-square wavefront tilt by noting that the deviation from a plane wave at any point x is given by

$$\Delta y = \frac{\lambda}{2\pi}\,\Delta\phi\,,$$

where y is in the direction of propagation. The angle associated with this deviation is

$$\theta = \frac{\Delta y}{x} = \frac{\lambda}{2\pi}\frac{\Delta\phi}{x}\,;\; \theta^2 = \frac{\lambda^2}{4\pi^2}\frac{\Delta\phi^2}{x^2} = \frac{\lambda^2}{4\pi^2}\frac{k_2}{\lambda^2}\cdot\frac{x^{5/3}}{x^2}\,;$$

therefore, the mean-square fluctuation is

$$\langle\theta^2\rangle = \frac{2}{D}\int_0^{D/2}\frac{\lambda^2}{4\pi^2}\cdot\frac{k_2}{\lambda^2}\,x^{-1/3}\,dx = \frac{2}{D}\int_0^{D/2}\frac{k_2}{4\pi^2}\,x^{-1/3}\,dx\,,$$

which, when solved, yields

$$\langle\theta^2\rangle = \frac{2\cdot 3\,k_2}{4\pi^2\cdot 2^{2/3}\cdot D^{1/3}}\,.$$

From Section 9.1, we may write k_2 in terms of the coherence diameter r_0 as

$$k_2 = \frac{32\cdot 2^{5/3}\lambda^2}{r_0^{5/}}\,;$$

the resultant root-mean-square angle of arrival fluctuation is

$$\theta_{\mathrm{rms}} = \sqrt{\langle \theta^2 \rangle} \approx \frac{\lambda}{r_0}\left(\frac{r_0}{D}\right)^{1/6} = \frac{\lambda}{D}\left(\frac{D}{r_0}\right)^{5/6} . \tag{9.2}$$

Thus we see that, in the heterodyne case, the detection efficiency decreases appreciably when the angle-of-arrival fluctuation approaches one beam width, which is what we would expect, intuitively. This second calculation also gives us information about incoherent detection, in the sense that the detector must now be large enough to take into account the angle-of-arrival fluctuations or, in the case of an aperture much larger than r_0, the blur circle that results from variations of the positions of the focal spot associated with separate regions of coherence diameter r_0. For a receiver with focal ratio $f\# = f/D$, the size of the spot on the detector for no turbulence is simply $f\#\lambda$, whereas for a coherence diameter r_0 the spot size is

$$d \approx f\# \cdot \lambda \cdot \left(\frac{D}{r_0}\right)^{5/6} .$$

Throughout these discussions, we have ignored amplitude fluctuations due to turbulence. These are distributed log-normally, i.e., the $\ln P$ fluctuates about a mean value $\overline{\ln P}$ with Gaussian distribution. Qualitatively, the power fluctuations decrease with increasing aperture, owing to increased averaging over the larger aperture; however, there is no simple relationship as in the phase-fluctuation case. *Fried* (1967b) treats the amplitude fluctuations for the heterodyne case.

Problems

9.1 The coherence diameter r_0 is a measure of the distance over which the optical phase is correlated. If there is a wind of velocity $v = 10\ \mathrm{m\ s^{-1}}$, transverse to the received beam path, estimate the spectral spread of the received signal due to turbulence at wavelengths of 1 and 10 μm under "good" seeing conditions. (Assume a fixed turbulence distortion that moves with the wind).

9.2 For a detection system pointed at the zenith, the majority of the turbulence distortion occurs within the first kilometer. Estimate the spectral spread as in problem 9.1, but for the receiver scanning at an angular rate of $1°\ \mathrm{s^{-1}}$.

10. Detection Statistics

Up to this point, we have described the efficiency of an optical or infrared detector in terms of its output signal-to-noise ratio, without questioning the relationship between this measure and the operating characteristics of the system. In this section, we shall treat the properties of the detection process that are particularly important for a radar or communications system. These are the probability of detection p_d and the probability of error or false alarm p_{fa}. In essence, we shall determine the effect of the signal-to-noise ratio upon the expectation of detecting a radar return or communication signal in the presence of noise, subject to an always finite probability that the supposed signal is really caused by noise in the detection system. One of the critical properties that affects this system behavior is the statistics of the received signal; we shall first treat the expected behavior of backscattered returns from a radar target in terms of the spectral width of the transmitted signal and the target properties. In this case, we shall find two extremes, namely a constant return power for any return pulse or an exponential probability distribution (Rayleigh) of pulse powers, that is, a mean-square fluctuation in received power that is equal to the square of the mean. We shall not attempt to treat the expected fluctuations in a communications channel, because these are difficult to specify, "fading" being caused by a large number of different phenomena such as turbulence, multipath scattering, and receiver-pointing fluctuations. Because the constant and the exponential statistics represent two extremes of fluctuation statistics, their treatment effectively encompasses the range of expected behaviors in radar and communication systems.

10.1 Statistics of Target Backscatter

For a simple target such as a single point scatterer, the reflected wave is an exact replica of the transmitted wave and we expect a power return that is independent of position or orientation of the target. Unfortunately, real targets are more apt to consist of a large set of individual scatterers and, at optical and infrared frequencies, the scatterers are all of about equal amplitude. There may be strong "glints" or specular reflections from individual flat segments, but the large number of scattering elements causes a typical target to approach so-called "diffuse" scattering, which is characterized by an ensemble of random fields, which result in complex Gaussian noise, just as in the case of thermal radiation. For this idealized model, we ask about the temporal statistics of the return power at the receiver in several different situations. First, we consider a diffuse target in the presence of a monochromatic wave. We also assume that the target is in the far field, that is, it is completely contained in one spatial mode of the receiver. In this case, the power statistics are Rayleigh, as given by

$$p(P) = \frac{1}{\bar{P}} e^{-P/\bar{P}},$$

where the statistics are taken over the ensemble of all possible orientations of the target. This results from the coherent addition of a set of monochromatic (or single-frequency) fields from all scattering elements, each field element having random amplitude and phase. Thus, if the target is nonrotating, a series of measurements will lead to a constant-power return but at an arbitrary level within the overall distribution. If the target rotates, the power fluctuates with the same distribution as the foregoing, with a *rate* determined by the following argument. If the scatterers at one side of the object move with respect to the opposite side by a distance $\lambda/2$, the addition of the scattered fields at the receiver produces an entirely independent power sample, because, on the average, the return from each scatterer has changed phase by 180°, because of the round-trip phase delay of 360° over the full target extent. The signal thus fluctuates with a characteristic period given by the time to rotate through an angle $\lambda/2D$, where D is the target dimension. Thus the fluctuation frequency of the power spectrum has a maximum value of $f = 1/T = (2D/\lambda)\, d\theta/dt$. Another way of thinking of this fluctuation is to note that the difference of Doppler shift between the two sides of the target is given by

$$\Delta f_{\mathrm{D}} = \frac{2v_1}{\lambda} - \frac{2v_2}{\lambda} = (2D/\lambda)(d\theta/dt),$$

which states that the frequency of the return signal is spread by Δf and that the power fluctuations thus have a maximum frequency content of Δf. This is identical with the behavior of thermal radiation with a spectral bandwidth $\Delta\nu$. We now return to our stationary target and ask how "monochromatic" must the radiation be to yield Rayleigh statistics. To treat this question, we assume a *constant-power* transmitted wave, but with finite spectral width. The spectral width may be determined by the pulsewidth or by the spectral content of the transmitter source. If the source is thermal, such as a flashlamp, then the *transmitted* radiation is already Rayleigh distributed and any target will exhibit an exponentially distributed return but with a fluctuation rate equal to the inverse of the linewidth. In the case of a laser source, the power transmitted is constant but may consist of a large number of frequencies spread over a finite spectral range that is determined by the type of laser. As an example, neodymium: YAG lasers produce a frequency band of the order of 10 GHz width, but the saturation of gain in the oscillator causes the total power to be fixed. In this case, if all of the random scatterers are at the same range, the return is Rayleigh distributed in an ensemble sense, just as in the case of the monochromatic wave. The reason is that each frequency component experiences the same time delay from each scatterer, which corresponds to a phase shift independent of frequency as long as the spectral spread is small compared with the optical frequency. Thus, the separate field components at each frequency are completely correlated and the return behaves as if the incident wave were monochromatic. If the target has depth or is tilted with respect to the line of sight, this criterion no longer holds and the fluctuations average out at any instant. Akin to the treatment for the rotating target, we ask over what depth or change of range does the relative phase shift of the scatterers change 180°, thus yielding an independent sample of the returned power. In this case, there is an average shift of half a wavelength across the frequency band if $\Delta\nu/\nu$ is equal to $\lambda/2d$, where d is the depth of the target along the line of sight. This distance,

$$d = \frac{\nu\lambda}{2\Delta\nu} = \frac{c}{2\Delta\nu},$$

is one-half of the coherence length of the incident radiation because of the round trip. We conclude that there will be as many independent samples of the power distribution in a single measurement as there are distances d in the depth of the target. In contrast, in the rotating target with monochromatic radiation, there are as many independent samples as there are time increments T in the actual measurement time. If we denote the number of independent samples as N, then we may write that the mean-square power fluctuation is

$$\overline{\Delta P^2} = N\,\overline{\Delta P_k^2} = N\,\bar{P}_k^2 = N\left(\frac{\bar{P}}{N}\right)^2 = \frac{1}{N}\,(P^2)\,, \tag{10.1}$$

where P_k is the average or expectation power for each independent sample. This is a simplified version of the central-limit theorem, which states that for any probability distribution, the final distribution for a large number of independent measurements approaches a Gaussian distribution about the mean of the original distribution with mean-square deviation equal to $1/N$ times the mean-square value of the initial distribution. Thus, if the target depth is much greater than the coherence length $c/\Delta\nu$, or the Doppler spread is much greater than the matched bandwidth of the receiver, the power fluctuation is small, and we treat the system in terms of a constant-power return. Otherwise, Rayleigh statistics apply and we use an exponential probability distribution.

The foregoing arguments hold for a target that is in a single spatial mode of the receiver. They therefore apply in all cases for coherent or heterodyne detection, because the receiver couples to only one mode of free space. In the case of incoherent detection, the criteria still apply, but only if the target is smaller than the diffraction-limited beam width of the receiver aperture. Otherwise, if the detector size is larger than the diffraction-limited focal spot, or equivalently, if the angular field of view is greater than the diffraction-limited field of view, then an extended target results in several independent samples of the reflected power within a single measurement. Just as in the derivation of Section 7.1, the fluctuations in each mode are independent and the net fluctuation for a diffuse target fills N diffraction-limited beam solid angles is given by (10.1). In closing, we note that these statistics explain the so-called "speckle" of a diffuse surface illuminated by a laser beam, that is, the eye or optical receiver resolves the scattered power into separate spatial elements, and the power is distributed among the individual elements in a Rayleigh manner.

10.2 Detection and False-Alarm Probability for the Incoherent Case

We first treat the case of a signal in the presence of "white", i.e., constant spectral density Gaussian, noise. This case applies to any incoherent detector whose output noise is determined either by circuit thermal noise or by "shot" noise with a large number of photoelectron events per measurement interval. This corresponds to the case in which the signal or background power is such that $r = P/h\nu$ is much greater than the detection bandwidth B, as in Section 2.2. We shall treat the photoelectron counting case in Section 10.4.

For both shot noise and thermal noise, the *ac* component of the current from the detection system is Gaussian distributed with zero mean and has a probability density given by

$$p(i) = \frac{1}{\sqrt{2\pi \overline{i_N^2}}} \exp(-\, i^2/2\overline{i_N^2}) \,.$$

In the presence of a signal current i_S, the comparable probability distributions will be as shown in Fig. 10.1. We now ask for the probabilities of false alarm and detection, if we set a threshold current i_T as the criterion for occurrence of a measureable event. We assume that the bandwidth B is matched to the spectral bandwidth of the signal, which in turn determines the value of $\overline{i_N^2}$ for the particular detecton. Shown in Fig. 10.1 are the appropriate areas that represent the detection and false-alarm probabilities. Once we find these probabilities, the false-alarm rate will be approximately $p_{fa}B$, because we are sampling the data at rate B. The integrals associated with the two probabilities are

$$p_{fa} = p(i_N > i_T) = \frac{1}{\sqrt{2\pi \overline{i_N^2}}} \int_{i_T}^{\infty} \exp(-\, i^2/2\overline{i_N^2})\, di \,,$$

$$p_d = p(i_S > i_T) = 1 - \frac{1}{\sqrt{2\pi \overline{i_N^2}}} \int_{(i_T - i_S)}^{\infty} \exp(-\, i^2/2\overline{i_N^2})\, di \,,$$

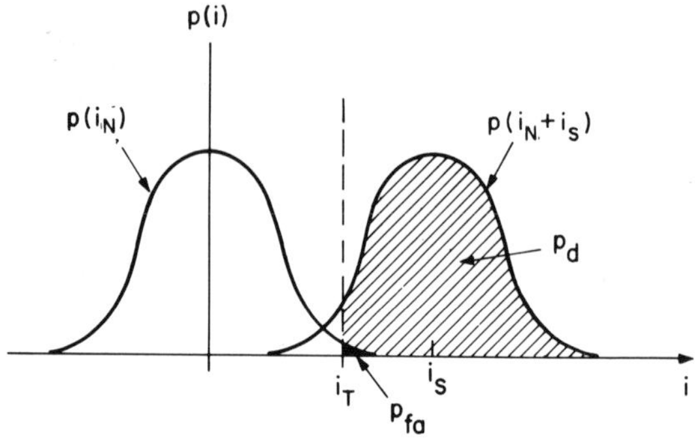

Fig. 10.1. Probability density for dc signal current in Gaussian noise

by reference to Fig. 10.1. The integrals are forms of the error function; the results, adapted from EOH (1974), are given in Fig. 10.2 We may examine the limiting case of low p_{fa} and high p_{d},

$$p_{\mathrm{fa}} = \frac{1}{\sqrt{2\pi \overline{i_N^2}}} \int_{i_T}^{\infty} \exp(-\, i^2/2\overline{i_N^2})$$

$$= \frac{1}{\sqrt{2\pi \overline{i_N^2}}} \exp(-\, i_T^2/2\overline{i_N^2}) \int_0^{\infty} \exp[-\,(x^2 2 i_T x)/2\overline{i_N^2}]\, dx\, .$$

For i_T^2 much greater than i_N^2, the last integral becomes

$$\int_0^{\infty} \exp[-\,(x^2 + 2 i_T x)/2\overline{i_N^2}] \approx \int_0^{\infty} \exp[-\,(i_T/\overline{i_N^2})x]\, dx \approx \frac{\overline{i_N^2}}{i_T}\, .$$

Thus, for low false-alarm probability,

$$p_{\mathrm{fa}} \longrightarrow \frac{\sqrt{\overline{i_N^2}}}{i_T \sqrt{2\pi}} \exp(-\, i_T^2/2\overline{i_N^2})\, .$$

As an example, for

$$\frac{i_T}{\sqrt{\overline{i_N^2}}} = 6\ ;\ p_{\mathrm{fa}} \approx 10^{-9}\, .$$

For the detection probability, we may use a similar approximation, yielding

$$p_{\mathrm{d}} \longrightarrow 1 - \frac{\sqrt{\overline{i_N^2}}}{(i_S - i_T)\sqrt{2\pi}} \exp[(i_S - i_T)2/2\overline{i_N^2}]\, ,$$

and for $(i_S - i_T)/\sqrt{\overline{i_N^2}} = 2$,

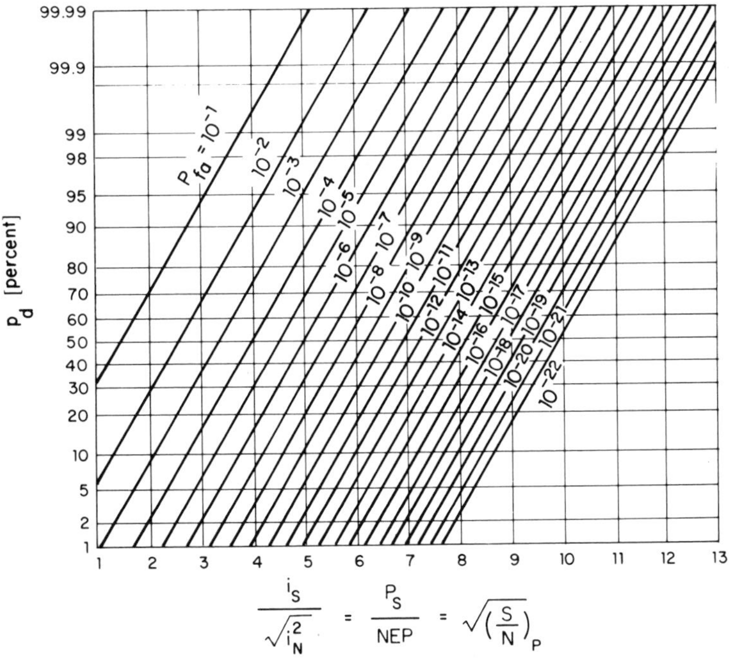

Fig. 10.2. Detection and false-alarm probabilities as a function of $(S/N)_V$ (after EOH, 1974)

$$\frac{i_S}{\sqrt{\overline{i_N^2}}} = \frac{P_S}{NEP} = 8 \; ; \quad P_D \approx 98\% \, .$$

We now consider the case of a signal that obeys Rayleigh statistics, i.e., the signal fluctuates with an exponential probability density from *pulse* to *pulse*. In this case, we must average the detection probability over the pulse-power distribution

$$p_d = \int_0^\infty p_d(i_S) p(i_S) \, di_S \; ; \quad p(i_S) = \frac{1}{\overline{i_S}} \, e^{-i_S/\overline{i_S}} \, .$$

This is a very complicated integral; it is obvious that the approximations never apply, because i_S always has values less than i_T. We can, however, approximate the result for $\overline{i_S}$ much greater than $\sqrt{\overline{i_N^2}}$, by ignoring the noise-current contribution. Specifically, we write

$$p_d = p(i_S > i_T) = \int_{i_T}^\infty \frac{1}{\overline{i_S}} \, e^{-i_S/\overline{i_S}} \, di_S = e^{-i_T/\overline{i_S}} \, ,$$

where the dominant term is the threshold current i_T. Thus, if i_T is much greater than the noise current, then the fluctuations of i due to the noise are small compared to i_T. If we define a threshold input optical power by P_T, the detection probability is

$$p_d = e^{-P_T/\overline{P_S}} \to \left(1 - \frac{P_T}{\overline{P_S}}\right); \quad P_S \gg P_T .$$

As an example of this case, consider our previous calculation, which yielded $p_{fa} = 10^{-9}$ and $p_d = 98\%$. In the Rayleigh-distributed case, with the same threshold ratio of 6, the p_{fa} remains the same, but the detection probability is

$$P_d = e^{-i_T/\overline{i_S}} = e^{-6/8} = 47\% .$$

Thus, the fluctuating-target statistics seriously degrade the detection probability, on a *single pulse* basis. Fortunately, most incoherent detection systems do not suffer from the results of pure Rayleigh statistics. As discussed in Section 10.1, either rapid fluctuations during the pulse, broad spectral bandwidth, or multiple spatial modes ameliorate the strong fluctuations. *Goodman* (1965) gives a general treatment of the fluctuating target in the incoherent-detection case.

10.3 The Coherent Case

The heterodyne or coherent-detection case is somewhat different than the previous one, because the *power* rather than the current out of the detector is proportional to input power. In addition, the signal current is at the i.f. frequency rather than being dc. In this case, we must "detect" the current for the second time by means of an envelope or square-law detector, in order to observe the waveform of the received signal. The best detection process for signal-to-noise ratios greater than unity is the envelope detector (*Davenport* and *Root*, 1958), which measures the peak value of the current averaged over many cycles of the center frequency. Such a device is usually a simple rectifying diode that drives a low-pass filter, matched to the signal-envelope waveform. The instantaneous envelope for band-limited Gaussian noise is exactly of the same form as that discussed in Section 7.1, where we are now considering the current rather than the electric field. The current is again complex Gaussian distributed because there are in-phase and out-of-phase components with respect to the center frequency, and each component is Gaussian. For the noise-current envelope alone, the probability density is

$$p(i) = \frac{i}{\overline{i_N^2}} \exp(-i^2/2\overline{i_N^2}) ,$$

whereas for the noise plus signal, the distribution is much more complicated,

$$p(i) = \frac{i}{\overline{i_N^2}} \exp[-(i^2 + i_S^2)/2\overline{i_N^2}] \cdot I_0\left(\frac{i i_S}{\overline{i_N^2}}\right),$$

where $I_0(z)$ is the modified Bessel function of zero order. Fig. 10.3 shows the equivalent distributions and their physical meaning in terms of the two current-envelope coefficients, i_1 and i_2, the in-phase and quadrature components.

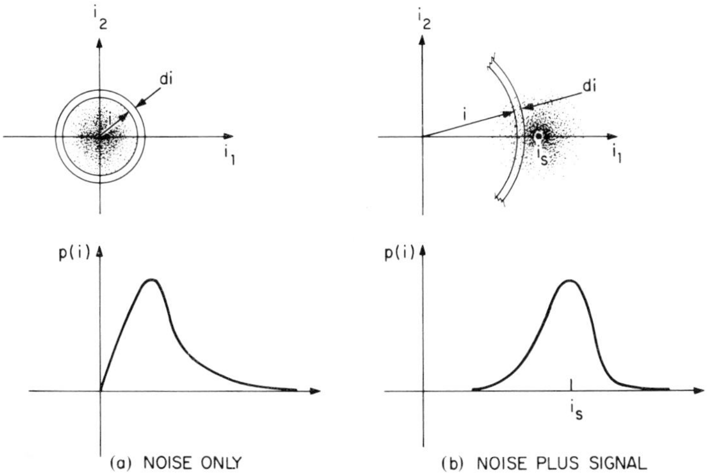

Fig. 10.3a and b. Probability density for envelope of sinusoidal signal in complex Gaussian noise

The mathematics of the p_{fa} and p_d calculation are similar to that of the incoherent detector, except more complicated because of the Bessel-function term. The results are widely used in microwave-radar theory and are discussed in detail by *Barton* (1964), including the case of a Rayleigh-distributed target return. The overall results are similar to the incoherent detector case except that one must take into account that i_S is proportional to $P_S^{1/2}$ in the heterodyne case. Thus, a signal-to-rms-noise-current ratio of 7 becomes an input signal-to-noise power of 7^2 or 49. Fig. 10.4 and 10.5, adapted from *Barton* (1964), give the results for constant and fluctuating targets.

10.4 Photoelectron-Counting Case

In the case of signal- or background-noise-limited detection, such as in the photomultiplier, detection may be carried out by counting the number of

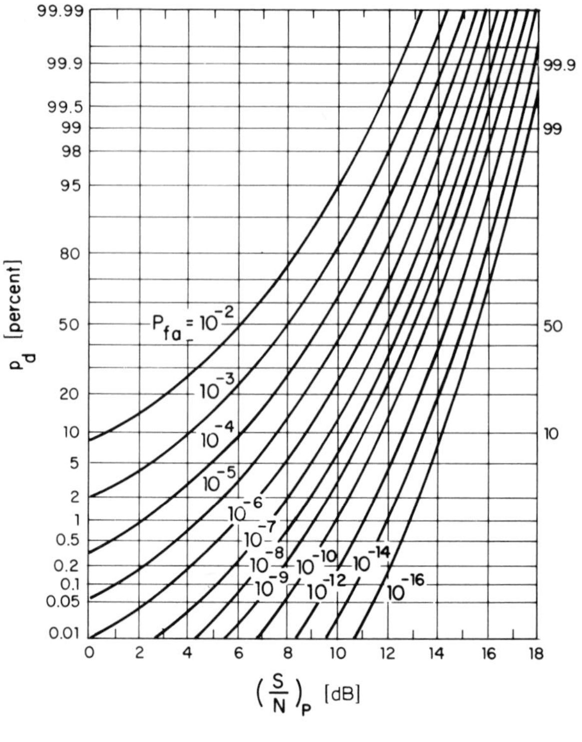

Fig. 10.4. Detection and false-alarm probabilities as a function of $(S/N)_P$ for a nonfluctuating target (after *Barton*, 1964)

individual pulses obtained during a measurement interval. For large counts per interval, the detection and false-alarm probabilities are given by the first treatment in Section 10.2. However, for a small number of events, say from 1 to 100, this detection technique yields different results, which we treat as follows. The probability of k pulses occurring in a measurement interval with average count of n is given by Poisson statistics as

$$p\,(k, n) = \frac{n^k e^{-n}}{k!}\,.$$

If we set a threshold count level k_T, p_{fa} and p_d may be written

$$p_{fa} = \sum_{k_T}^{\infty} p(k,n) = \sum_{k_T}^{\infty} \frac{(n_B)^k\, e^{-n_B}}{k!}\,,$$

$$p_d = \sum_{k_T}^{\infty} p(k,n) = \sum_{k_T}^{\infty} \frac{(n_B + n_S)^k\, e^{-(n_B+n_S)}}{k!}\,,$$

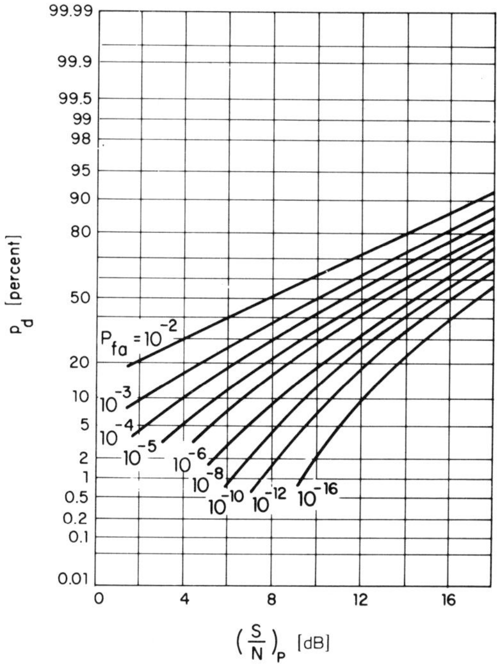

Fig. 10.5. Detection and false-alarm probabilities as a function of $(S/N)_P$ for target with Rayleigh fluctuation (after *Barton*, 1964)

where n_B is the average background and dark-current count, and n_S is the average signal count. These expressions have been evaluated and may be found in EOH (1974). We can consider one interesting case analytically, the case of zero background or dark current, in which case we set $k_T = 1$, and note that the false-alarm rate is zero, because $n_B = 0$. This corresponds to the signal-noise-limited case. The resultant detection probability is

$$p_d = \sum_1^{\infty} p(k,n) = \left[\sum_0^{\infty} p(k,n)\right] - p(0,n) = 1 - e^{-n},$$

which applies for constant amplitude of the return pulses. If the pulses are Rayleigh distributed, we average the detection probability over the pulse-height distribution,

$$\begin{aligned} p_d &= \int_0^{\infty} p_d(n)\, p(n) dn = \int_0^{\infty} (1 - e^{-n}) \frac{1}{\bar{n}} e^{-n/\bar{n}}\, dn \\ &= \left(1 - \frac{1}{1+\bar{n}}\right) = \frac{\bar{n}}{\bar{n}+1} \longrightarrow 1 - \frac{1}{\bar{n}}\ ;\ \bar{n} \gg 1\,. \end{aligned}$$

In terms of the actual incident power, we may set $\bar{n} = \eta P_S \tau / h\nu$; setting B, the post-detector bandwidth, $= 1/2\tau$, we obtain $\bar{n} = \eta P_S/2h\nu B$. But $2h\nu B/\eta$ is the *NEP* for a signal-noise-limited photon detector and the final result, for $\bar{n}$ large is

$$p_d \longrightarrow \left(1 - \frac{1}{\bar{n}}\right) = \left(1 - \frac{NEP}{P_S}\right),$$

which is identical with the expression previously derived for the Gaussian-noise case, if we identify the threshold power with the *NEP*, which is indeed the case for zero background count.

Problems

10.1 A diffuse spherical satellite target of diameter 1 m travels at 8 km s^{-1} at an altitude of 300 km. Using single-frequency illumination at 10 μm wavelength, what is the approximate fluctuation rate of the Rayleigh-distributed return as the target passes directly overhead?

10.2 The same target is observed with a Nd:YAG laser with spectral linewidth 10 GHz. Assume that the target is *not* resolved by the receiver aperture and find the fractional root-mean-square fluctuation of the power returned.

10.3 How large an aperture would be required to obtain the same smoothing, assuming single-frequency operation and a detector with unlimited field of view?

11. Selected Applications

Applying the results of the previous chapters, we conclude with several examples of detector use in measurement systems. Common to most of the applications is the question of the optimum detection process, coherent or incoherent. The answer depends upon the wavelength, the spectral bandwidth, and the relative merits of sensitivity versus system complexity. Although we do not consider thermal detectors as such in this material, their use is obvious in cases where the ultimate in sensitivity is not essential. The final section on intensity interferometry is of special interest since it combines in one experiment several of the fundamental concepts developed in the text. These include the thermal-radiation field fluctuations of Chapter 7 and the target-scatter properties discussed in Chapter 10.

11.1 Radar

As an example of detector use in radar systems, we shall consider a short-range system that operates at either of two of the common laser wavelengths, 1.06 μm (Nd:YAG) or 10.6 μm (CO_2). For comparison of operation at the two wavelengths, we specify the range R to be 10 km, the aperture diameter d, 10 cm, and require a range resolution of 150 m, corresponding to a pulse width of 1 μs. We also assume a near-horizontal path along the ground and a radar-target cross section σ of 10^{-2} m^2. We first determine the received power at the detector solving the radar equation in a step-by-step process. Specifically, we first calculate the intensity of the optical radiation at the target as

$$I_{\mathrm{T}} = \frac{P_{\mathrm{T}}}{\Omega R^2}\,\mathrm{e}^{-\alpha R}\,,$$

where P_{T} is the peak transmitted power, Ω is the beam solid angle, and α is the atmospheric-attenuation coefficient. The power received by the aperture is then

$$P_{\mathrm{S}} = \left(\frac{\pi d^2}{4}\right)\cdot\frac{\sigma I_{\mathrm{T}}}{4\pi R^2}\,\mathrm{e}^{-\alpha R} = \frac{\sigma d^2 \mathrm{e}^{-2\alpha R}}{16\Omega R^4}\cdot P_T\,,$$

where we use the standard radar definition of cross section σ, which is the area that scatters the incident radiation *isotropically*, that is, into 4π steradians, and produces the appropriate power density at the receiver. The ratio of signal power to transmitted power, or the "path loss", is then

$$P_S/P_T = \frac{\sigma d^2 e^{-2\alpha R}}{16 \Omega R^4} = \frac{6 \times 10^{-22}\, e^{-2\alpha R}}{\Omega}\,.$$

We must now consider the solid angle Ω and the absorption coefficient α. In the case of 1.06 μm radiation, the transmitted-beam solid angle is *not* λ/d, where d is the aperture diameter, but is λ/r_0, the coherence diameter as determined by turbulence because r_0 is less than d for the particular case considered. Specifically at 1.06 μm the coherence diameter for a 10 km ground path is typically 2 cm (*Gilmartin* and *Holtz*, 1974). The transmitter aperture thus behaves as a set of individual 2 cm diameter coherent transmitters and the beam solid angle is therefore $(\lambda/r_0)^2 = 2.5 \times 10^{-9}$ steradians. In contrast, the coherence diameter at 10.6 μm is much larger than d and the solid angle is the diffraction-limited value $(\lambda/d)^2 = 10^{-8}$ steradians, larger than the 1.06 μm value because of the longer wavelength, despite the lack of turbulence degradation. Taking into account the appropriate absorption coefficients for a horizontal path (EOH, 1974),

$$\alpha = 0.12\ \mathrm{km}^{-1} : \ 1.06\ \mu\mathrm{m}$$

$$\alpha = 0.3\ \mathrm{km}^{-1} : \ 10.6\ \mu\mathrm{m}\,,$$

we obtain for the overall path-loss ratios,

$$P_S/P_T = 2.2 \times 10^{-14}\ \text{at}\ 1.06\ \mu\mathrm{m}$$

$$P_S/P_T = 1.5 \times 10^{-16}\ \text{at}\ 10.6\ \mu\mathrm{m}\,.$$

We now examine the required peak power P_T as determined by the detector as well as by the desired detection and false-alarm probabilities.

For generality we shall consider both incoherent and coherent detection for either wavelength and determine which type results in the lowest required transmitter power. First in the 1.06 μm case, we note that a photomultiplier operating in the signal-noise-limited mode has an *NEP* of $2\Gamma h\nu B/\eta$. For a quantum efficiency of 1% now available at this wavelength and a Γ of 1.2 the resultant *NEP* for a bandwidth of 1 MHz is

$$NEP = \frac{2\Gamma h\nu B}{\eta} = 4 \times 10^{-11}\ \mathrm{W}\,.$$

A silicon avalanche photodiode would also yield an equivalent performance; the greater quantum efficiency is counteracted by the lower available gain before amplification. We must now be sure that signal-noise-limited operation is feasible, by calculating the background-limited *NEP*,

$$(NEP)_{BL} = \sqrt{\frac{2h\nu B P_B}{\eta}}\,.$$

The background in the worst case would be radiation scattered from a sunlit cloud, whose irradiance (*Wolfe,* 1965) is 10^4 μWcm^{-2} ster-μm^{-1}. For the solid angle of reception 2.5×10^{-9} steradians, aperture area, 78 cm^2, and a 1% optical bandwidth filter, P_B becomes 2×10^{-11} W. The resultant background-limited *NEP* is therefore

$$(NEP)_{BL} = \sqrt{(4 \times 10^{-11})(2 \times 10^{-11})} = 2.8 \times 10^{-11}\ \mathrm{W}\,,$$

so that the detector is essentially signal-noise limited, using a readily available narrowband filter.

If we consider coherent or heterodyne detection for this system, we are faced with two difficulties. First, the atmospheric-coherence diameter limits the effective aperture to 2 cm, rather than the full 10 cm. The effective received power is thus reduced by $(2/10)^2$ or is 4% of the value calculated from the radar equation. This would still yield comparable sensitivity to the incoherent photomultiplier, because heterodyne detection could utilize a high-quantum-efficiency photoconductor or photodiode. Unfortunately, the spectral width of Nd:YAG laser radiation is of the order of 10 GHz; the resultant bandwidth B in the heterodyne minimum-detectable power $h\nu B/\eta$ is thus 10^4 times the value used in the incoherent case. By techniques similar to those discussed under radiometry, in the next section, we can perform some "post-detection" integration and regain a factor $(B\tau)^{1/2}$, where τ is the pulsewidth, but this gain of 100 still places the sensitivity two orders of magnitude below that of the incoherent system. In addition, the requirement of a local-oscillator source adds unneeded complication to the system.

For operation at 10.6 μm, we first examine incoherent detection and choose the most-sensitive-possible device, the photoconductor, because the semiconductor photodiode is limited to high-background use by its low leakage resistance. From Section 6.1, a typical photoconductor may have a D^* given by

$$D^* = 2.7 \times 10^{16} \sqrt{1/f_c} = 2.7 \times 10^{13}$$

in the absence of background radiation, where we have taken f_c as 1 MHz, corresponding to the 1 μs pulsewidth. This corresponds to an *NEP*, as a function of area, of

$$NEP = \sqrt{AB}/D^* = 4 \times 10^{-11} \sqrt{A}\ ;$$

if we choose a detector dimension of 0.1 mm, then the *NEP* becomes 4×10^{-13} W. If we consider the background, we may refer to Fig. 2.4, and note that for 0.1 mm detector size, the width of the detector is equal to $(\lambda/d)f$, where d is the aperture size and f is the focal length. In this case, d/f becomes 0.1 and $\sin(\theta/2)$ for the detector reception angle is sin (0.05 rad) = 0.05. Thus, D^* for unit emissivity background at 300 K is $5 \times 10^{10}/[0.05. (2)^{1/2}] = 7 \times 10^{11}$; the square root of two comes from the photoconductive mode of operation. Because this background-limited D^* is approximately 1/40 of the amplifier-noise-limited D^*, then the scene emissivity and the reduction of background by a narrowband infrared filter would have to result in a 1600-fold net reduction of background. These requirements are well beyond the state of the art; consequently, the *NEP* of such an incoherent detector is approximately 10^{-12} W or greater.

In contrast, heterodyne detection is quite feasible at this wavelength, because the coherence diameter is much larger than the aperture and the spectral linewidth for a CO_2 laser can be much less than 1 MHz, the matched-receiver bandwidth. Assuming a semiconductor photodiode with 50% quantum efficiency the minimum detectable power is $h\nu B/\eta = 4 \times 10^{-14}$ W. Although a separate local oscillator is required, the heterodyne mode offers the advantage of simultaneous Doppler velocity measurement, with a single-pulse resolution of 5 m s^{-1}, corresponding to the i.f. bandwidth of 1 MHz.

To summarize the system requirements to this point, for $S/N = 1$, the optimum detectors for the two wavelength regions require the following peak transmitter powers:

Wavelength [μm]	Detector power [W]	Path loss	Transmitter power [W]
1.06	4×10^{-11}	2.2×10^{-14}	1800
10.6	4×10^{-14}	1.5×10^{-16}	270

The energy per pulse is thus 1.8 and 0.27 mJ, respectively.

Although the 10.6 μm system seems, at first, to be the more efficient, consideration of false-alarm and detection probabilities contradicts that conclusion. In the 1.06 μm case, the spectral width of the laser source, 10 GHz, results in a coherence "depth" at the target of $c/2\Delta\nu$ or 1.5 cm. Similarly, if the target size is much greater than $\lambda R/d = 10$ cm, the target return will have small fluctuation and we may use a constant-amplitude return. From the previous section, a p_{fa} of 10^{-6} with $p_d = 98\%$ requires a $(S/N)_V$ of 7/1. The required peak power is thus 12.5 kW, or 12.5 mJ/pulse. In contrast, even for a nonfluctuating target return in the 10.6 μm case, the required output $(S/N)_P$, which is the same as the input $(S/N)_P$, is 14 dB or 25/1, as determined from Fig. 10.4 for the same p_{fa} and p_d. Thus, the required transmitter peak power is 6.8 kW or 6.8 mJ/pulse. Taking into account fluctuations, because the coherence "depth" is equal to the range resolution of 150 m, because the spectral width is narrow, the required $(S/N)_P$ from Fig. 10.5, again for $p_{fa} = 10^{-6}$ and $p_d = 98\%$, becomes 29 dB, or 800/1! Fortunately, unless the target and transmitter are both stationary, relative rotation of the target will cause fluctuations on a pulse-to-pulse basis and the high transmitter power is not essential, if we are willing to wait for several pulses for assured target detection.

11.2 Radiometry and Spectroscopy

We now treat the general topics of radiometry and spectroscopy, where we wish to measure either the incident power in a finite wavelength band or, in the case of spectroscopy, the variation of power as a function of wavelength. We may be observing a self-emitting source or we may be measuring the variation of transmission of a medium illuminated by a blackbody source such as the sun or by a laser in a long-path transmission measurement of atmospheric pollutants. In any event, we are not interested in the phase of the optical or infrared wave, only in the net power incident upon the detector. Despite this fact, we shall find that coherent or heterodyne detection is superior for *some* applications, specifically when the desired resolution or spectral bandwidth is small.

We first consider the incoherent case that uses the measuring system shown in Fig. 11.1, which is a switching radiometer. The switch may be a tunable optical filter (or a spectrometer) which alternates between two different wavelengths. In the latter case, for a small wavelength shift, the system output measures the derivative of the source spectrum. A switch of some kind is required, because constant input power produces dc output, which fluctuates with

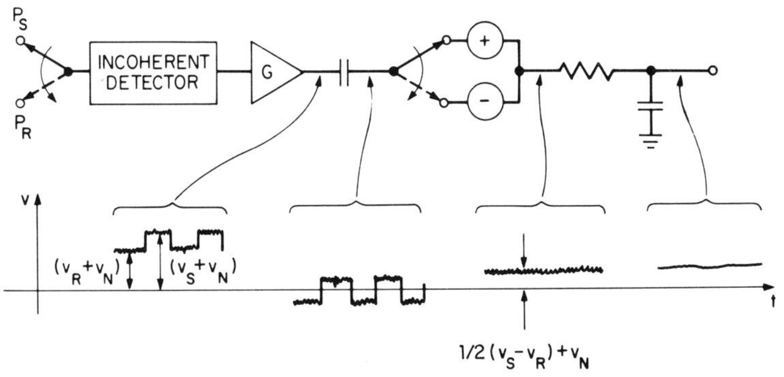

Fig. 11.1. Switching radiometer. Signal and reference voltages exaggerated for clarity

amplifier gain as well as being degraded by low-frequency additive noise in the detector and amplifier. Shown schematically in the figure are the different waveforms throughout the signal path. The switch at the right of the circuit reverses the polarity of the incoming current, at a rate synchronized with the input switch; thus the final dc current out of the low-pass filter is one-half the difference between the source- and reference-induced currents. Note also that we have described the low-pass filter by an RC filter with time constant $\tau = RC$. The switching frequency f_s is well above the frequency at which excess amplifier and detector noise is significant, typically 100 to 1000 Hz, yet low enough so that detector frequency response is not limiting. With these considerations we may write the measured output voltage as

$$v_{\mathrm{OUT}} = (1/2)\,(v_{\mathrm{S}} - v_{\mathrm{R}}) = (K/2)(P_{\mathrm{S}} - P_{\mathrm{R}})\,,$$

where the constant K includes the responsivity of the detector and the gain of the amplifier. The factor of one-half comes from the 50% duty cycle involved in the switching process. In a similar manner, we may write the mean-square flucutation of the output voltage as

$$v_{\mathrm{N}} = K(NEP) = K(NEP)_{1\ \mathrm{Hz}} \cdot B^{1/2} = \frac{K(NEP)_{1\ \mathrm{Hz}}}{\sqrt{4\tau}}\,,$$

because the effective bandwidth of an RC filter is $1/4\tau$. Note that there is no factor of one-half, as in v_{out}, because alternating the sign of the noise voltage at the second switch has no effect upon the mean-square noise voltage, noise being a random process.

The final ratio of signal voltage to root-mean-square voltage fluctuation or $(S/N)_{\mathrm{V}}$ is given by

$$\left(\frac{S}{N}\right)_{\mathrm{V}} = \frac{1}{2}\,\frac{(P_{\mathrm{S}} - P_{\mathrm{R}})\,\sqrt{4\tau}}{(NEP)_{1\mathrm{Hz}}} = \frac{(P_{\mathrm{S}} - P_{\mathrm{R}})\,\sqrt{4}}{(NEP)_{1\mathrm{Hz}}}\,.$$

We shall examine this relationship in more detail after treatment of the coherent detector in radiometry and spectroscopy.

The appropriate circuit for coherent detection is shown in Fig. 11.2, with appropriate waveforms. In contrast to the incoherent radiometer, the output of the heterodyne detector is a band-limited rf signal rather than a dc term proportional to input power. As a result, the measurement of the source properties requires detection of this broadband power by a second detector. Because the signal-to-noise ratio at the input to the detector is generally much less than unity, a square-law detector is the optimum device (*Davenport* and *Root*, 1958) for converting the net power to a measurable voltage. The circuit is similar to that for incoherent radiometry in requiring synchronous switching. The arguments that follow are also directly applicable to a system that uses laser preamplification followed by an incoherent optical or infrared detector; the latter replaces the electrical square-law detector in the figure. In any event, a critical aspect of this type of detection is the spectral bandwidth of the measured radiation, which not only determines the power, as in the incoherent case, but also directly affects the fluctuation rate at the input to the square-law device. In treating this quality of the coherent system, we shall use B, the i.f. bandwidth, interchangeably with $\Delta\nu$, the spectral bandwith, because the heterodyne-conversion process produces a one-to-one correspondence between the two quantities. As we shall also see, the fluctuation of the output from the square-

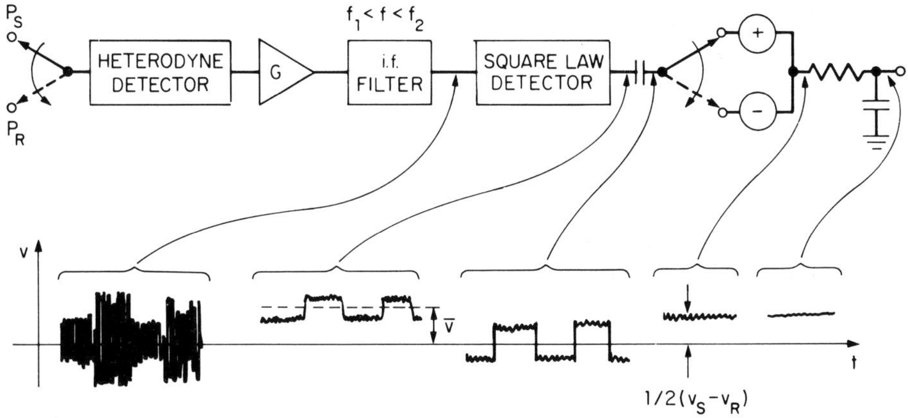

Fig. 11.2. Heterodyne switching radiometer. Signal and reference voltages exaggerated for clarity

law detector has exactly the same properties as those deduced previously in Section 7.1 for the fluctuations associated with the classical thermal-radiation field as processed by an incoherent photodetector. Thus, we shall implicitly justify our statement regarding the replacement of the heterodyne-square-law-detector combination by a laser-preamplifier-incoherent-detector chain.

For generality, we shall treat the output noise or voltage fluctuation for the case when finite-bandwidth constant-spectral-density noise occurs between frequencies f_1 and f_2, where $(f_2 - f_1)$ is defined as bandwidth B. The mean-square-noise spectral density is plotted in Fig. 11. 3a; the noise spectrum at the output of the square-law detector is shown in Fig. 11. 3b. The low-frequency noise spectrum that slopes to zero at $B = (f_2 - f_1)$ results from the difference-frequency beating of the individual frequency contributions in the input-noise band. This is identical with the behavior as shown in Fig. 7.1 in our treatment of the radiation-field fluctuations. The second triangular plot represents the

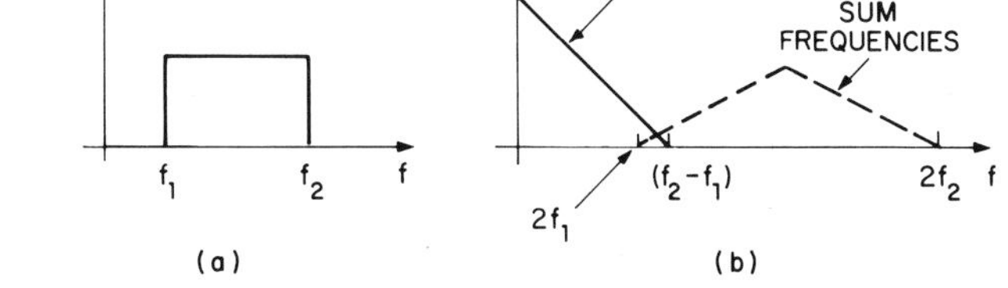

Fig. 11.3. (a) Input-noise current to square-law detector. (b) Output-noise voltage spectra

sum-frequency mixing across the noise band, which extends from $2f_1$ to $2f_2$. Because we are simply convolving the input spectra with itself, either in a sum or difference operation, the integrated noise power under each curve is the same. Because the low-frequency term is the one of direct interest, as it introduces noise that is passed by the low-pass post-detection filter, we could use the results of Section 7.1 directly. In other words, the low-frequency noise voltage has an exponential probability distribution, as previously, and the total mean-square voltage fluctuation is the square of the mean voltage, i.e., the dc voltage out of the detector. We shall prove this in a slightly different manner, this time by examining the statistical behavior of the complete noise voltage at the square-law-detector output, as follows. We first write the appropriate description of the probability distribution of the input current as

$$p(i) = \frac{1}{\sqrt{2\pi \overline{i_N^2}}} \exp(-\, i^2/2\overline{i_N^2}) \, .$$

This is single-variable Gaussian distributed, because it describes the instantaneous value of the current, rather than the envelope of the two quadrature components of the center frequency $(f_2 - f_1)/2$. The square-law detector operates on this current according to the expression $v = ki^2$, and the resultant probability density for the output voltage may be shown to be

$$p(v) = \frac{1}{\sqrt{2\pi k \overline{i_N^2}\, v}} \exp(-\, v/2k\overline{i_N^2})$$

by setting $p(v)\, dv = p(i) di$, with $dv = 2\, kidi$. The first and second moments of the distribution are

$$\bar{v} = \int_0^\infty vp(v)dv = \frac{1}{\sqrt{2\pi k \overline{i_N^2}}} \int_0^\infty v^{1/2} \exp(-\, v/2k\overline{i_N^2})\, dv = k\overline{i_N^2}\, ,$$

$$\overline{v^2} = \int_0^\infty v^2 p(v)dv = \frac{1}{\sqrt{2\pi k \overline{i_N^2}}} \int_0^\infty v^{3/2} \exp(-\, v/2k\overline{i_N^2})\, dv = 3(k\overline{i_N^2})^2\, ,$$

where the integrals have been evaluated by use of the gamma function and its properties,

$$\Gamma(p) = \int_0^\infty x^{(p-1)}\, e^{-x}\, dx$$

$$\Gamma\left(\frac{3}{2}\right) = \frac{\sqrt{\pi}}{2}\, ; \, \Gamma\left(\frac{5}{2}\right) = \frac{3\sqrt{\pi}}{4}\, .$$

As we would expect, the mean voltage $\bar{v}$ is simply $k\overline{i_N^2}$, whereas the mean-square fluctuation is given by

$$\overline{\Delta v^2} = \overline{v^2} - \bar{v}^2 = 2\bar{v}^2 .$$

Here we note that the fluctuation is twice the square of the mean; but then we note that half of that fluctuation is in the sum-frequency spectrum and the other half in the difference-frequency component, based upon our previous discussion. Thus, we have shown in a different manner the validity of our earlier treatment. We may calculate the noise-voltage spectral density at frequencies much less than B by writing

$$(\overline{\Delta v^2})_{\text{low-freq.}} = \bar{v}^2 = \int_0^B \overline{\Delta v^2}(0)\left(1 - \frac{f}{B}\right) df$$

$$\therefore \quad \overline{\Delta v^2}\,(0) = \frac{2\bar{v}^2}{B} .$$

If we pass this noise-voltage spectrum through a low-pass RC filter with time constant τ, the resultant root-mean-square voltage fluctuation is

$$v_{\text{rms}} = \sqrt{\overline{\Delta v^2}(0) \cdot \frac{1}{4\tau}} = \frac{\bar{v}}{\sqrt{2B\tau}} ,$$

as long as the low-pass filter bandwidth is much less than B. A physical understanding of this expression may be obtained by considering that $B\tau$ independent samples of the current are being taken, and that the fluctuation of the measure is the square root of the number of samples. Many texts arrive at the expression $(B\tau)^{1/2}$ without the factor of 2. This arises because the latter form applies to a true integrator, which takes a running sum over time τ. The effective sampling time τ for various ideal and real filters is discussed by *Bracewell* (1965), who gives a more elegant derivation of the noise behavior.

We now determine the signal-to-noise voltage at the output of the system, in the same manner as in the previous case. The signal voltage is proportional to one-half the difference between the source power and the reference power, whereas the mean voltage due to noise has the same proportionality to the minimum detectible power, $h\nu B/\eta$. Thus, the $(\mathrm{S/N})_{\mathrm{V}}$ is

$$\left(\frac{S}{N}\right)_{\mathrm{V}} = \frac{1}{2}\frac{\eta(P_{\mathrm{S}} - P_{\mathrm{R}})}{h\nu B}\sqrt{2B\tau} = \frac{\eta(P_{\mathrm{S}} - P_{\mathrm{R}})}{h\nu} \cdot \sqrt{\frac{\tau}{2B}} .$$

We note that the spectral bandwidth has a direct bearing on the ultimate signal-to-noise ratio of the system. It is interesting to compare the results for the incoherent and coherent systems:

Radiometric detection		
Incoherent		Coherent
$(S/N)_V = \dfrac{(P_S - P_R)}{2(NEP)}$		$(S/N)_V = \dfrac{\eta(P_S - P_R)}{h\nu} \cdot \sqrt{\dfrac{\tau}{2B}}$
Signal-noise limited	$\dfrac{\eta(P_S - P_R)\tau}{h\nu}$	
Background limited	$\dfrac{\sqrt{\eta}(P_S - P_R)\sqrt{\tau}}{\sqrt{2h\nu P_B}}$	
Amplifier limited	$\dfrac{(P_S - P_R)\sqrt{\tau}}{(NEP)_{1\,Hz}}$	

Here we note that incoherent detection is always superior in the signal-noise-limited case, unless the low-pass-filter time constant is comparable with the inverse spectral bandwidth, a nonrealistic case. In the background-limited case, we reach the interesting conclusion that the two systems are equivalent if the background "count" $P_B/h\nu$ is equal to the spectral bandwidth being measured. This may be seen by rewriting the sensitivity as

$$\left(\frac{S}{N}\right)_V = \frac{\sqrt{\eta}(P_S - P_R)}{h\nu}\sqrt{\frac{\tau}{2P_B/h\nu}}$$

and assuming that η is close to unity. No generalizations may be made about the amplifier-limited case, other than comparing the two expressions to obtain the relative sensitivities. These results are not surprising if we recall that a heterodyne or coherent system measures the phase as well as the amplitude of the wave; the process of measurement introduces extra noise, as in the uncertainty principle. Basically, if we wish to measure only the power or photon flux, a coherent system overmeasures by also extracting the phase, the resultant information being discarded after final filtering. Despite this shortcoming, there are many applications for which the desired spectral resolution is narrow enough, so that coherent detection is superior either because of a high background or the nonavailability of a signal-noise-limited device. In any comparison, of course, it

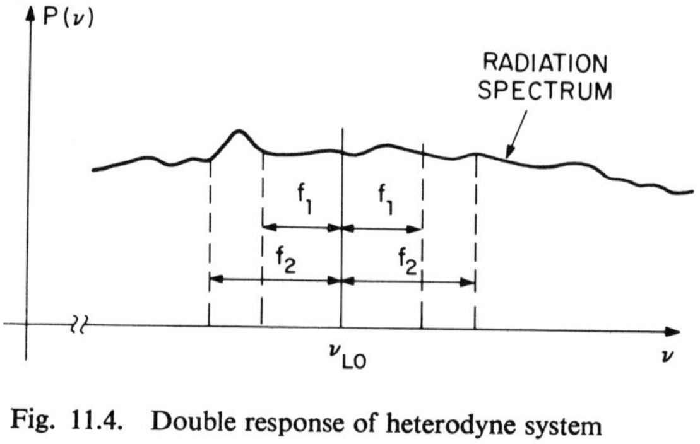

Fig. 11.4. Double response of heterodyne system

should be remembered that the source power, at least for radiometric measurements, is proportional to spectral bandwidth for a broad-spectrum source; in this case, the incoherent system usually excels.

As a final note, we point out that the heterodyne system detects power in a bandwidth both above and below the local-oscillator frequency, as depicted in Fig. 11.4. Therefore, unless there is a narrowband optical filter before the detector, the source and reference power should be calculated using *twice* the i.f. bandwidth as the spectral width. For spectroscopy, this "image effect" may be ignored or can be deleterious, depending upon the complexity of the spectral shape. These arguments do not apply in the case of laser preamplification, because the laser determines the specific passband for the system.

11.3 Stellar Interferometry

We first consider amplitude interferometry, as first carried out by Michelson and Pease. In this technique, the interference between two wave-fronts from the distant source is used to determine the angular subtense of the star. Later, we shall consider intensity interferometry, which depends upon the correlation of the wave powers rather than their amplitudes and phases. Consider two receiving mirrors, as shown in Fig. 11.5. Here we have limited the size of the mirrors to a diameter r_0, the coherence diameter determined by the atmospheric turbulence. The beams from each mirror are combined at a detector; because this mixing of the beams is similar to that in the heterodyne case, it does not help to increase

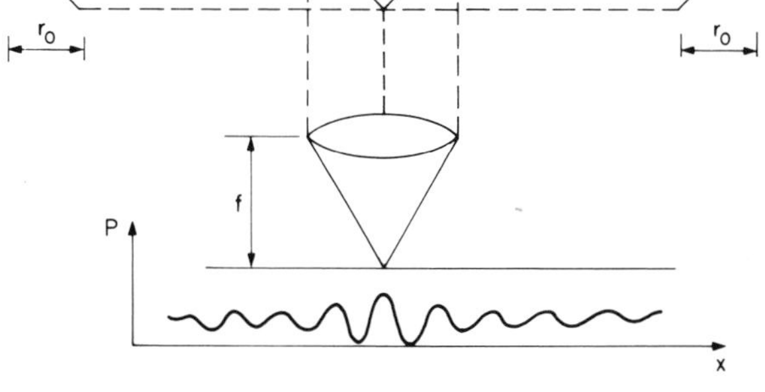

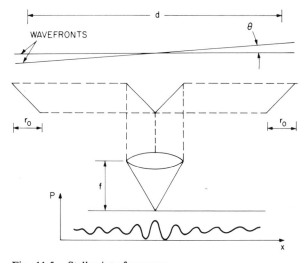

Fig. 11.5. Stellar interferometer

the sizes of the mirrors; the mixing product reaches a maximum at the size r_0. The atmosphere also introduces tilts into the wavefront; as a result the average wavefront will fluctuate over an angle of the order of 1 arc-s thus causing the pattern at the focal plane to shift back and forth in the x direction. If the power that strikes each aperture is P_S, it may be shown that the pattern for a point source at infinity is as given below, where the displacement in x corresponds to a change of angle, $\theta = x/f$:

$$P(x) = 2\frac{dP_S}{d\nu}\Delta\nu\left[1 + \frac{\sin\left(\frac{\pi\Delta\nu xd}{fc}\right)}{\left(\frac{\pi\Delta\nu xd}{fc}\right)}\cos\frac{2\pi\nu_0 xd}{fc}\right],$$

where $\Delta\nu$ is the spectral linewidth of the radiation and $dP/d\nu$ is the spectral-power density. The fringe visibility, or contrast, which may be defined as the (average-minimum)/average thus remains close to unity for small arguments of the sin x/x term in the brackets. Because the fringe spacing is $x = f\lambda/d$, the number of fringes that have near unity visibility is approximately $\nu/\Delta\nu$. We now ask what happens if the source is finite and the radiation subtends an angle $\Delta\theta$, the subtense of the star. In this case, the power may be shown to be

$$P = 2\frac{dP_S}{d\nu}\Delta\nu\left[1 + \frac{\sin\left(\frac{\pi d}{\lambda}\Delta\theta\right)}{\left(\frac{\pi d}{\lambda}\Delta\theta\right)}\cos\frac{2\pi xd}{f\lambda}\right],$$

if $\Delta\nu/\nu$ is assumed to be small enough so that the angular spread determines the fringe contrast, rather than the finite spectral bandwidth. Thus for a non-turbulent medium, we may determine the subtense of the star by increasing d until the fringe visibility drops to a small value, say 10%. Michelson and Pease observed this effect visually; by waiting for times when the angle fluctuation was small or varying very slowly, they could estimate the fringe visibilities and determine the size of the star. In more modern techniques, two photon detectors might be placed at a spacing that corresponds to adjacent maximum and minimum $P(x)$ values; by switching from one to the other, we might measure the average fringe visibility, provided that the switching time is rapid compared with the atmospheric fluctuations. Actually, to collect all of the energy, we need a set of photon detectors spaced so as to correspond to all of the maxima and minima in the fringe pattern, which spreads over the diffraction-limited image due to the aperture r_0. Thus, approximately d/r_0 detectors are necessary. With a single pair of detectors the signal power would be only $(r_0/d)P_S$. Assuming the ideal case of a full set of detectors, we must limit the spectral bandwidth of the signal, because the fringe pattern will move across the detectors to the extent of (d/r_0) fringes, associated with the angle fluctuations. Therefore, the spectral linewidth is limited to $\Delta\nu/\nu = r_0/d$. Assuming ideal detection, such as by a photomultiplier, we set the minimum detectable power $h\nu/\tau$ equal to one-tenth of the received power, because this will give 10 percent accuracy of measurement of total power, or a minimum-detectable fringe visibility of 10%. The result is

$$\frac{h\nu}{\tau} = 0.1 P_S = 0.1 \, f h\nu\Delta\nu \left[\frac{(\Delta\theta)^2 r_0^2}{\lambda^2}\right] = 0.1 \, f h\nu \left(\frac{r_0}{d}\nu\right)\frac{(\Delta\theta)^2 r_0^2}{\lambda^2},$$

where f is the photon occupancy factor and the last fraction is the ratio of the star-subtense solid angle to the diffraction-limited solid angle for the receiver aperture. Solving for $\Delta\theta$ yields

$$\Delta\theta = \sqrt{\frac{10\lambda^2 d}{f\nu r_0^3 \tau}} = \frac{\lambda}{r_0}\sqrt{\left(\frac{d}{r_0}\right)\frac{10}{f\nu\tau}} = \frac{\lambda}{d}\sqrt{\left(\frac{d}{r_0}\right)^3 \frac{10}{f\nu\tau}}$$

from which we can estimate the integration time required to make the observation. Using the star, Sirius, with $T = 10{,}000$ K, and angular subtense of 0.007 arc-s $= 3.5 \times 10^{-8}$ radians, we find a value of $d = 14$ m for the fringe visibility to drop to zero. Using $r_0 = 5$ cm for typical atmospheric seeing, $f = 0.1$ for the given temperature, and a wavelength of 0.5 µm, we find an integration time of approximately 10 µs for this ideal case. If we had used just a single pair of detectors, the required integration time would increase to d/r_0 times as great or 280×10 µs $= 2.8$ ms. We might have used heterodyne detection, except that the spectral bandwidth of $6 \times 10^{14}/280 = 2 \times 10^{12}$ would require a frequency response of the detector well beyond present capabilities and the minimum detectable power would increase from $h\nu/\tau$ to $h\nu\sqrt{B/\tau}$. Thus, the required integration time would increase by the ratio $\sqrt{B\tau}$ or 4×10^3 times that required in the multiple-detector case.

In the 10 µm region, the use of heterodyne detection is somewhat more attractive, especially because the increased baseline required for equivalent resolution requires extremely long path lengths to combine the beams from the two apertures. The larger value of r_0 would allow aperture sizes of the order of a meter, but the optical isolation and combination of such beams would be exceedingly difficult. In the example, if we assume the same value of d/r_0, *i.e.*, $d = 280$ m for $\Delta\theta = 0.007$ arc-s, a star (infrared) at 500 K, and an r_0 of 1 m, the required integration time is found to be 3 s with a spectral bandwidth of 10^{11} Hz. For the incoherent case,

$$\Delta\theta = \frac{\lambda}{d}\sqrt{\left(\frac{d}{r_0}\right)^3 \cdot \frac{0}{f\nu} \cdot \frac{NEP}{h\nu\sqrt{\tau}}},$$

where the ideal-photodiode minimum detectable power has been replaced by the expression for the amplifier-noise-limited case

$$P_{\min} = \frac{(NEP)_{1\,\mathrm{Hz}}}{\sqrt{\tau}} ,$$

and we have assumed an NEP of 10^{-15} for the detector. The minimum power is thus

$$P_{\min} = \frac{10^{-15}}{\sqrt{\tau}} = \frac{10^{-15}}{\sqrt{3}} \approx 6 \times 10^{-16}\ \mathrm{W} .$$

For comparison, if we had used heterodyne detection with unit quantum efficiency, the value of $P_{\min}$ would be

$$P_{\min} = h\nu \sqrt{\frac{B}{\tau}} = 2 \times 10^{-20} \sqrt{\frac{10^{11}}{3}} = 6 \times 10^{-15}\ \mathrm{W} ,$$

where we have used $B = 10^{11}$ Hz. As may be seen, the sacrifice of sensitivity is rather small; it would require about a 3-fold increase of integration time. Unfortunately, present detectors are limited to a B of 10^9 Hz, which would require an additional 10-fold increase of integration time, to compensate for the reduced power input in the smaller spectral bandwidth. The advantage of heterodyne detection at each telescope receiver, by use of a small laser beam transmitted to each unit makes it attractive, but probably for shorter baselines than 280 m, or larger star subtenses.

11.4 Intensity Interferometry

Hanbury Brown and *Twiss* (1956; 1957a,b) used radiation-field fluctuations to do intensity interferometery, based upon the fact that the fluctuations of light from a distant source are correlated over an angle given by λ/d', where d' is the size of the source. This is equivalent to the case of the rotating target in Section 10.1, where a change of angle of $\lambda/2d'$ produced uncorrelated fluctuations. The factor of two arises from the two-way path in the radar case. It is easily shown that this means that the intensity fluctuations, that is, the *second* term in the noise-current expression (7.3), are completely correlated for a detector spacing of zero, and the correlation disappears when the spacing d is such that λ/d approximates the subtense of the target. The correlation coefficient may be shown to be the square of the Fourier transform (*Born* and *Wolf*, 1964, p. 510) of the radiance distribution over the source. Or put another way, the angular distribution of the correlation function on the ground is the square of the Fraunhofer transform of the source distribution. We now take two detectors, in this case perfect photon counters (in the *Hanbury Brown* and *Twiss* experiment, photomultipliers), and calculate the voltage out of a square-law detector that receives both currents, as shown in Fig. 11.6. The mean-square fluctuation of the currents is given by

$$\overline{i_1^2} = \overline{i_{1S}^2} + \overline{i_{1C}^2} = \frac{2e^2\eta P_1 B}{h\nu} + \frac{2e^2\eta P_1 B}{h\nu}\,\eta f_{\mathrm{eff}}$$

$$\overline{i_2^2} = \overline{i_{2S}^2} + \overline{i_{2C}^2} = \frac{2e^2\eta P_2 B}{h\nu} + \frac{2e^2\eta P_2 B}{h\nu}\,\eta f_{\mathrm{eff}}\,,$$

and

$$\overline{i_1^2} = \overline{i_2^2}\ ;\ \overline{i_{1S}^2} = \overline{i_{2S}^2}\ ;\ \overline{i_{1C}^2} = \overline{i_{2C}^2}\ \text{since}\ P_1 = P_2\,.$$

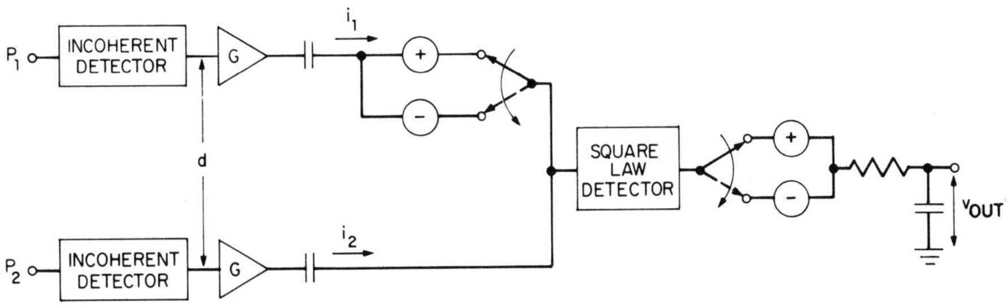

Fig. 11.6. Intensity interferometer

Calculating the voltage output, we obtain

$$\begin{aligned}\bar{v}_{\mathrm{OUT}} &= k\,\overline{i^2} = k\,(\overline{i_1^2} + \overline{i_2^2} + \overline{2i_1 i_2}) = 2k\overline{i_1^2} + 2k(\overline{i_{1S}i_{2S}} + \overline{i_{1C}i_{2S}} + \overline{i_{2S}i_{1C}} + \overline{i_{1C}i_{2C}})\\ &= 2k\overline{i_1^2} + 2k\overline{i_{1C}i_{2C}}\,,\end{aligned}$$

because only i_{1C} and i_{2C} are correlated. The fluctuation of voltage, from the radiometer expressions, is

$$v_{\mathrm{rms}} = \frac{2k\,\overline{i_1^2}}{\sqrt{2B\tau}}$$

and the signal-to-noise ratio is

$$\frac{v_S}{v_{\mathrm{rms}}} = \frac{\overline{i_{1C}^2}}{\overline{i_1^2}} \times \sqrt{2B\tau} = \eta f_{\mathrm{eff}}\,\sqrt{2B\tau}$$

for complete correlation. Note here that the switching, which is performed at a rate high compared with atmospheric fluctuations, causes the sign of the $i_{1C}i_{2C}$ to change at the switching rate; the resulting measured voltage is the value of this correlation.

In the experiment by *Hanbury Brown* and *Twiss*, the collector apertures were 1.5 m and the source was Sirius, with $T = 10{,}000$ K and a subtense of 0.007 arc-s, or 3.5×10^{-8} radians. From these quantities, we may calculate $\mathscr{f}$ to be 0.1 at 0.5 μm, and $\mathscr{f}_{\text{eff}}$ becomes 10^{-3}, because the ratio of the star solid angle to the receiver-beamwidth solid angle is 10^{-2}. With a quantum efficiency of 20% and a detector bandwidth of 40 MHz, the signal-to-noise ratio is unity for an integration time of 0.3 s. For measurement of the point at which the correlation drops to near zero at the appropriate detector spacing, a signal-to-noise ratio of about ten would be adequate; this would require an integration time of 30 s. In the actual experiment, the required integration times were much longer, because of excess noise in amplifiers and recording systems.

Problems

11.1 A heterodyne receiver is used to observe emission and absorption lines from a planet at temperature T, with average emissivity ε. If the planet is resolved by the receiver aperture, derive an expression for the $(S/N)_V$ as a function of the desired resolution $\Delta\nu$, the integration time τ, and the fractional change of emission (or absorption) α, due to the characteristic line. Neglect atmospheric effects.

11.2 For $\lambda = 10$ μm, $T = 300$ K, $\varepsilon = 1$, find the limiting resolution $\Delta\nu$ for an integration time 1 s, a fractional emissivity change of 0.01, and a $(S/N)_V = 5$.

References

Barton, D. K. (1964): *Radar System Analysis* (Prentice-Hall, Englewood Cliffs)
Born, M., Wolf, E. (1964): *Principle of Optics*, 2nd revised ed. (Pergamon Press, Oxford)
Boyd, G. D., Gordon, J. P. (1961): *Bell Syst. Tech. J.* **40,** 489
Bracewell, R. N. (1965): *The Fourier Transform and Its Applications* (McGraw-Hill, New York)
Davenport, W. B., Root, W. L. (1958): *Random Signals and Noise* (McGraw-Hill, New York)
Emmons, R. B. (1967): J. Appl. Phys. **38,** 3705
EOH. (1974): *RCA Electro-Optics Handbook* (RCA/Commercial Engineering, Harrison, New Jersey)
Fante, R. (1975): Proc. IEEE **63,** 1669
Forrester, A. T., Gudmundsen, R. A., Johnson, P. O. (1955): Phys. Rev. **99,** 1691
Fried, D. (1967a): Proc. IEEE **63,** 57
Fried, D. (1967b): IEEE J. Quantum Electron. QE-**3,** 213
Gilmartin, T. J., Holtz, J. Z. (1974): Appl. Opt. **13,** 1906
Golay, M. J. E. (1949): Rev. Sci. Instr. **20,** 816
Goodman, J. W. (1965): Proc. IEEE **53,** 1688
Hanbury Brown, R., Twiss, R. Q. (1956): Nature **178,** 1046
Hanbury Brown, R., Twiss, R. Q. (1957a): Proc. Roy. Soc., A **242,** 300
Hanbury Brown, R., Twiss, R. Q. (1957b): Proc. Roy. Soc., A **243,** 291
Hudson, R. D., Hudson, J. W. (eds.) (1975): *Infrared Detectors* (Dowden, Hutchinson, and Ross, Stroudsburg)
Jones, R. C. (1959): Proc. Inst. of Radio Eng. **47,** 1481. Reprinted in: *Infrared Detectors*, ed. by R. D. Hudson J. W. Hudson (Dowden, Hutchinson, and Ross, Stroudsburg 1975)
Keyes, R. J. (ed.) (1977): *Optical and Infrared Detectors*, Topics in Applied Physics, Vol. 19 (Springer, Berlin, Heidelberg, New York)
Kruse, P. W. (1977): "The Photon Detection Process", in *Optical and Infrared Detectors*, ed. by R. J. Keyes, Topics in Applied Physics, Vol. 19 (Springer, Berlin, Heidelberg, New York)
Low, F. J. (1961): J. Opt. Soc. America. **51,** 1300. Reprinted in: *Infrared Detectors*, ed. by R. D. Hudson J. W. Hudson (Dowden, Hutchinson, and Ross, Stroudsburg 1975)
McIntyre, R. J. (1966): IEEE Trans. ED-**13,** 164
Melchior, H. M., Fisher, M. B., Arams, F. R. (1970): Proc. IEEE **58,** 1466. Reprinted in: *Infrared Detectors*, ed. by R. D. Hudson and J. W. Hudson (Dowden, Hutchinson, and Ross, Stroudsburg 1975)
Milton, A. F. (1977): "Transfer Devices for Infrared Imaging" in *Optical and Infrared Detectors*, ed. by R. J. Keyes, Topics in Applied Physics, Vol. 19 (Springer, Berlin, Heidelberg, New York)
Putley, E. H. (1970): "The Pyroelectric Detector" in *Semiconductors and Semimetals;* ed. by R. K. Willardson and A. C. Beer (Academic Press, New York)
Putley, E. H. (1977): "Thermal Detectors" in *Optical and Infrared Detectors*, ed. by R. J. Keyes, Topics in Applied Physics, Vol. 19 (Springer, Berlin, Heidelberg, New York)
Reif, F. (1965): *Statistical and Thermal Physics* (McGraw-Hill, New York)
Ross, M. (1966): *Laser Receivers* (John Wiley and Sons, New York)
Ross, A. H. M. (1970): Proc. IEEE **58,** 1766
Siegman, A. E. (1966): Proc. IEEE **54,** 1350
Smith, R. A., Jones, F. E., Chasmar, R. P. (1968): *The Detection and Measurement of Infra-Red Radiation*, 2nd ed. (Oxford-Press, London)
Spears, D. L. (1977): Infrared Phys. **17,** 5
Teich, M. C. (1969): Appl. Phys. Lett. **14,** 201
van der Ziel, A. (1954): *Noise* (Prentice-Hall, New York)

Wolfe, W. L. (ed.) (1965): *Handbook of Military Infrared Technology* (Office of Naval Research, Department of the Navy, Washington, D. C.)
Yariv, A. (1971): *Introduction to Optical Electronics* (Holt, Rinehart and Winston, New York)
Yariv, A. (1975): *Quantum Electronics*, 2nd ed. (John Wiley and Sons, New York)

Subject Index